Sarah Kuhn

Copyright © 1976 by Monthly Review Press
All Rights Reserved

Library of Congress Cataloging in Publication Data
Main entry under title
Technology, the labor process, and the working class
    Includes bibliographical references.
    1. Labor and laboring classes—addresses, essays, lectures. 2. Marxian economists—addresses, essays, lectures. 3. Labor and laboring classes—United States—addresses, essays, lectures. I. Baxandall, Rosalyn.
HD4854.T37   331   76-26122
ISBN O-85345-397-7

First Printing

Monthly Review Press
62 West 14th Street, New York, N.Y. 10011
21 Theobalds Road, London WC1X 8SL

Manufactured in the United States of America

Contents

The Working Class Has Two Sexes
*Rosalyn Baxandall, Elizabeth Ewen,
and Linda Gordon* ... 1

Work and Consciousness
John and Barbara Ehrenreich ... 10

Capitalist Efficiency and Socialist Efficiency
David M. Gordon ... 19

Division of Labor in the Computer Field
Joan Greenbaum ... 40

Marx as a Student of Technology
Nathan Rosenberg ... 56

Social Relations of Production and
Consumption in the Human Service Occupations
Gelvin Stevenson ... 78

The Other Side of the Paycheck:
Monopoly Capital and the Structure of Consumption
Batya Weinbaum and Amy Bridges ... 88

Marx versus Smith on the Division of Labor
Donald D. Weiss ... 104

Two Comments
Harry Braverman ... 119

Contributors

ROSALYN BAXANDALL teaches in the American Studies department at the State University of New York (SUNY), College at Old Westbury. She is co-author, with Linda Gordon, of *America's Working Women: A Documentary History of Working-Class Women in the United States*. ELIZABETH EWEN teaches in the American Studies department at SUNY, College at Old Westbury. She is currently working on a book about immigrant women in New York City at the turn of the century. LINDA GORDON is associate professor of history at the University of Massachusetts, Boston. She is the author of *Woman's Body, Woman's Rights* (September 1976), a book on the history of birth control. JOHN EHRENREICH teaches in the American Studies department at SUNY, College at Old Westbury. He and Barbara Ehrenreich are co-authors of *The American Health Empire: Power, Profits, Politics*. BARBARA EHRENREICH has written numerous article and pamphlets. She is co-author, with Deirdre English, of *Witches, Midwives, and Nurses: A History of Women Healers*. DAVID M. GORDON teaches economics at the New School for Social Research. He is a member of the Union for Radical Political Economics. JOAN GREENBAUM, a former computer programmer, now teaches data processing at LaGuardia Community College in the Bronx. NATHAN ROSENBERG is a professor of economics at Stanford University. GELVIN STEVENSON, a member of the Union for Radical Political Economics, has worked for both the City and State of New York. He is currently doing community organizing in the Bronx. BATYA WEINBAUM is a member of the Women's Work Project of the Union for Radical Political Economics. AMY BRIDGES is on the editorial board of *Politics and Society*. She is currently at the University of Chicago. DONALD D. WEISS teaches philisophy at SUNY, Binghamton. HARRY BRAVERMAN is the author of *Labor and Monopoly Capital: The Degradation of Work in the Twentieth Century* and is the director of Monthly Review Press.

The Working Class Has Two Sexes
by Rosalyn Baxandall, Elizabeth Ewen,
and Linda Gordon

From a feminist perspective, Harry Braverman's focus in *Labor and Monopoly Capital*[1] on the way industrial capitalism reorganizes and continually splits up the labor process has special importance. This focus provides the basis for a renewed revolutionary view of people's labor, an insistence that socialism cannot be created by a mere transfer of ownership of products but must transform the processes of work and daily life. The powerlessness of women and all working-class people is not based solely on material deprivation but on the stunting of our human capacities by an oppressive division of labor.

Yet to understand the extent of the damage created by the division of labor, Braverman's distinction between the social division of labor and the detailed division of labor in capitalist industry is not adequate. (The former refers to gross occupational distinctions, as between carpenters and printers, farmers and merchants; the latter refers to the splitting up of labor processes within single production units as with assembly line work or when a garment worker stitches one particular seam only.) Braverman argues in Chapter 3 that the social division of labor before the introduction of capitalist commodity production, did not create fundamental inequality: "While the social division of labor subdivides society, the detailed division of labor subdivides *humans,* and while the subdivision of society may enhance the individual and the species, the subdivision of the individual . . . is a crime against the person and against humanity."

It seems to us that the social division of labor has also been a crime against the person. We must begin by recalling a society in which the labor of production and reproduction did not divide itself, in time or in location; only capitalism, through the establishment of the factory and the sale of labor power, created that separation. Nevertheless, long before capitalism or industrialism a *hierarchical* division of labor arose which secured for the male sex the more prestigious, remunerative, and powerful work—whose specifics changed as the mode of production shifted from hunting and gathering to agriculture and crafts—forcing women into the most confining, least respected work. From this division of labor flowed a division of the human character into feminine and masculine qualities which in turn diminished the breadth and complexity of intellectual and emotional capacity in men and women.

Thus capitalism based its division of labor upon a preexisting, culture-wide division of labor which had already created gross inequality and fractured human beings. The craftsmen and farmers whose labor Braverman shows to have been more skilled and whole before capitalism already enjoyed privileges resting on the degradation of female labor. Peasants and artisans alike had wives who fed and clothed them, raised their children, and cleaned up after them. This degradation of female labor was continued and intensified by capitalism in unpaid as well as paid work. Furthermore, capitalism continued the fracturing of human beings in many areas not commonly considered work at all—in recreation, in reproduction and the socialization of children, in consumption. The actual experiences of commoditization and loss of control over one's purposes extend far beyond places of employment. The consciousness created by this alienation is formed not only in the ten or so hours a day that most people spend in wage labor, but also into the remaining hours of eating, buying, cleaning, cooking, gossiping, riding in cars, watching television; some of these consciousness-forming experiences may even be recapitulated, and impressed more deeply, in our dreams.

Thus by focusing exclusively on wage labor, Braverman has already ignored one of the most pernicious aspects of the division of labor. Furthermore, ignoring the unpaid work and activity of the working class impoverishes Braverman's analysis of the consciousness of wage laborers themselves. The consciousness formed outside the workplace is brought into the workplace where it helps to reinforce and to give a rational appearance to the irrationalities of degraded labor. It is true that

Braverman's book has gained in depth and precision because of its limited focus on wage labor and production processes, and it may be that his choice to limit it thus enabled him to make the maximum possible contribution. But in the long run a total view of the working class, and especially of its strategic power, requires viewing the whole lives of all its members.

Furthermore, just as life's experiences are of a piece for working-class people, so they are for capitalists. As a class, capitalists in the monopoly era must be concerned with production and realization as equally problematic and inseparable. As rulers they must not only squeeze surplus value out of workers but also coerce them to purchase the products of their own labor and to reproduce themselves in a manner suitable for the reproduction of the capitalist mode of production. In this unified set of tasks the capitalists need women and men, wage and non-wage workers, equally.

Women's Wage Labor

While industrialization changed considerably the nature of women's work, it also utilized an earlier historical sexual division of labor. A system of institutional sexism dominates work and social relationships both at jobs and at home. In fact, in order to understand this institutionalized system we must see the mutually reinforcing relationship of home and job, and the special position of women in the paid labor force. Much of the paid work that women do is an extension of their family work: serving, cleaning, supporting, restoring, caring. One-third of all women work in seven job categories: clerical, domestic, retail sales, waitressing, bookkeeping, nursing, teaching. Throughout the labor force women are "sexegated" into a separate labor market with different, less desirable jobs and lower pay.

In the era of competitive capitalism this "sexegation" was supported by the illusion that women were not breadwinners. Recent social history research is eroding this myth, for it turns out that working-class men's wages were rarely sufficient to support families and that women's earnings in the nineteenth century were characteristically essential, not supplementary. It is true, however, that in the pre-monopoly era it was not the norm for married women to work regularly outside their homes, and when they did so it was usually seen as a misfortune. Wives normally contributed to their family incomes through work done at home—taking in boarders, doing gardening, laundry, sewing, piece work, and a myriad of other kinds of paid work in their homes.

Today, in the era of monopoly capital, out-of-the-home employment of married women is becoming a norm, just as new labor demands have drastically increased the service and clerical sectors of the labor force. Braverman's book explores working conditions with the understanding that the whole working class is being transformed by these changes, that women predominate in the expanding sectors of the wage labor force, and that they may become a central part of the labor force. An important part of Braverman's thesis is that clerical and service work is becoming more factory-like. While we agree with this basic perception, we see important mediations, particularly as a consequence of the sexual division of labor. Sexism frequently presents an obstacle to the rationalization of work. Male executives treat their secretaries as wives on the job and expect them to cater to whims that are not exclusively job-defined. Thus executives object when their personal secretaries are removed to typing pools. There has been enormous resistance to the assembly-line reorganization of office work. ITT recently conducted a survey of their facilities to find out whether there had been a shift toward greater use of stenographic dictating machines, and found none. IBM has encountered resistance to the introduction of work processing both from secretaries and executives.

The helping nature of much of women's work also has special consequences. Much of women's wage work involves catering to the public, helping people, doing jobs which in a socialist society might contain positive humanistic value. Many of the services that women provide, unlike products, are not readily apparent to the public or themselves, and it is therefore harder for service workers to develop a sense of their own exploitation. It is also more difficult for service workers to strike, since they are aware that poor people may be severely harmed by their withdrawal of labor. On the other hand, because such women workers often identify with the customers and consumers they service, it may be easier for them in the long run to develop a class consciousness which encompasses the whole working class.*

Even if women are employed full-time and even if they enjoy and identify with their wage work, society and most women in it still define themselves mainly by their family roles. This female consciousness permeates the working day of employed women, often in the form of home obligations which may physically or mentally keep them away from their jobs, or in the way their bosses underestimate and misinterpret their behavior. Because of this dual work, women must make choices about how to bridge the home and work gap, and these choices often entail taking certain kinds of jobs which, though lower-paying and with less opportunity for improvement, keep them closer to their family role, provide a chance to meet men, allow them to dress prettily, for example. Thus it is impossible even to understand the workings of the labor market, the "public sphere," without looking at how the home and neighborhood, the "private sphere," affect it.

Women's Unpaid Labor

In the pre-monopoly era, women's activity in the home and community stood, in many ways, outside the domain of industrial capital. Though the wage system was increasingly molding the context of social life for working-class families, the family itself was not yet primarily defined by the rhythms and relationships of industrial production. The family remained, for most people, within a discreet world of old-world, peasant, and working-class culture. While women were dominated by patriarchal authority, their work in producing and reproducing their daily lives was under their own control. Even within the patriarchal context, men did not interfere in the making of food and clothing, family education, health care, and women's par-

---

* Another trend in service work is the elimination of many jobs, such as supermarket cashiers and saleswomen, due to computerization of functions. The degree and effects of these complex trends remain unpredictable.

ticipation in their communities. Within this sphere women practiced a form of universal, non-specialized, craftswomanship (from knitting, for example, to birth control) which was passed down generationally and shared within each generation among the members of their community.

With the rise of monopolies, new forms of social organization began to appear. The expanding commodity market of products and services effected a historic break in the relationship between women and industry. It is possible that monopoly capital was more decisive for the lives of most working-class women than the rise of capitalism itself. The need for controlled markets demanded a mobilization of all social resources for potential profit. Daily life itself was incorporated into the hegemonic schemes of capitalist domination; its terrain was being invaded, reassembled, and redefined.

The industrial and service sectors appropriated the traditional tasks of women and returned them in the form of finished commodities and institutionalized services. The social nature and meaning of women's work were being unrecognizably altered. The "degraded" labor process had direct ramifications within the "non-productive" sphere of the home. Women's work was changed to such an extent that the new conception of "housework" (a conception closely linked to monopoly capital and created by the first decades of monopoly rule) bore little resemblance to the universal craftsmanship of the past. Women's work had been changed from creating to caretaking. What women had once produced from start to finish was now being reconstituted as consumable, socially produced goods and services. As the home was reorganized into an internal market for capitalist penetration, the actual work of women was transformed from artisan craftsmanship to a definition not unlike "machine tending."

While machinery and modern products have indeed eliminated some of the traditional drudgery of women's work, they have been utilized within a context which encroached upon and severely regimented the social space that women inhabit. The technological invasion of the home is closely linked with the degradation of women's work. As goods and services were being created outside the home, the authority on how to use them moved out of the hands of the community and into the offices of corporate management. The basic impetus of scientific management—the separation between conception and execution—was increasingly informing women's work in the home. Capitalism was telling women how to work as well as what to buy for their daily needs; the home was becoming a model for corporate priorities. The labor process involved in the creation and recreation of daily life, from domestic hygiene to child care, passed from networks of women to the managers and their emissaries: doctors, principals, psychologists, marketing researchers. The contours of these processes were centralized and linked to the corporate chain of command. Women were now expected to perform home tasks which were defined and specifically directed by capital and promulgated by the media of advertising and other related consciousness industries. The changing standards of cleaning, child care, and sexuality are intimately linked to this development. In this new mode the home mirrored the work processes, time sequences, and hierarchical modes of authority that constituted the social dynamic of monopoly capital. Women were losing the ability to oversee their own work and were becoming subservient to new authorities which profit from and encourage the degradation of daily life itself.

Monopoly capital completed the penetration of daily life by expanding industries designed to lower consciousness and make women dependent on commoditized visions of social life. This as a life style allowed for the easy interchangeability of women from home to industry. For example, the directions on cake-mix boxes and shampoo (open bottle, wet hair, work in shampoo, rinse, repeat) have a remarkable similarity to the directions which Braverman quotes from industrial and office work. In both, the productive process remains a mystery for authorized personnel only. The goal of capitalist industry in relation to women is to place women firmly within the grip of capital; not only as finely tuned consumers, but also as workers in the home, ready for integration outside the home.

At the same time the "non-productive" industries train and develop a capitalistic women's consciousness: they produce values and images designed to prepare women to accept "degraded" labor in the home as well as in the work force. The passively defined woman inside the home correlates to a passive woman worker outside the home. The corporate attempt to monopolize home life is integral to the corporate attempt to define the labor process itself.

Conclusion

None of these comments is meant to diminish the importance of *Labor and Monopoly Capital*; we are responding to it because it is so important, and because it makes a major contribution, perhaps unbeknownst to its author, to feminist analysis. Nor are our comments intended to raise merely academic objections. They come primarily from strategic considerations concerning the building of a socialist feminist revolutionary movement. One of Braverman's biggest contributions is that he has understood the complexity of the contemporary working class, and the unity within its differences; he has not been content to pay lip service to the understanding that women are workers too, but has fully incorporated this understanding into his analysis. Without bearing in mind at all times the importance of both wage and non-wage labor in maintaining capitalism and creating the experience of the working class, socialists may fall into the error of equating the working class with workers. Precisely because the working class is a class, and not merely a collection of individuals, its culture and experience must be understood as simultaneously diverse and united. Women and men particularly exchange understandings and experiences, since they so often live together. Socialist strategies that do not take into consideration the totality of this experience run the risk not only of leaving out women, blacks, and other groups who have not historically been identified as the proletarian archetype, but equally run the risk of failing to address the actual experience of even white male workers.

Our thinking about these issues is of course primarily indebted to the women's liberation movement. Seeking to explain their own situation, feminist socialists have insisted on the totality of capitalist penetration of our lives and the possibility and necessity of total resistance to that system. Thus interpreted,

Braverman's findings become perhaps less pessimistic than they might at first appear. The debasement of skill and the degradation of labor have affected nearly all working-class people, including children and housewives; and this universality has created conditions for unity in a working class until now deeply and often hostilely divided within itself. Without this unity, we don't have a chance.

Notes

1. *Labor and Monopoly Capital* (New York, 1974).

Work and Consciousness
by John and Barbara Ehrenreich

There are some books—a very few—that explain so much that they are most likely to be criticized for not explaining *everything*. *Labor and Monopoly Capital* falls in this category; it can be faulted mainly for what it does not do, for failing to answer the questions it never raised in the first place. The most frustrating "omission" of this sort is the one which Paul Sweezy acknowledges in the book's foreword: the failure to analyze "the *subjective* aspects of the development of the working class under monopoly capitalism." (our emphasis) Yet we think that *Labor and Monopoly Capital* has a lot to say about the "subjective" issues which preoccupy socialist activists. Braverman does not give the answers (after all, he never raised the questions), but we will argue that he gives some pretty broad hints.

The problem is this: Braverman argues with exceptional thoroughness that *objectively* monopoly capitalism is following the course foreseen by Marx. The working class has grown to include some three-quarters of the work force; the division of society into two classes—the proletariat and the bourgeoisie—is almost complete. Braverman describes the objective development of the working class: the destruction of individual craftsmanship among industrial workers, the "proletarianization" of large sectors of the work force outside of the traditional industrial proletariat, the continuing "accumulation of misery" within the class. But he has nothing to say about the other part of Marx's vision: the development of the working class as the conscious agent of socialist revolution. Why didn't the U.S. working class become a *class-for-itself* (to use Marx's language) at the same time as it developed as a *class-in-itself*?

Marxists (and Marx himself) have tended to assume that the same forces which produced the proletariat as a class-in-itself would, all other things being equal, generate proletarian class *consciousness*. At the same time as the process of capital accumulation impoverishes the workers and levels their individual skills, it brings them together in factories and other large socialized workplaces. In these settings the workers can grasp the contradiction between the collective nature of production and the private ownership of the means of production. The productive process itself should serve to demonstrate the collective might of the workers and (perhaps with the help of a few agitational leaflets) the parasitical nature of the bourgeoisie. The fact that this has not yet happened is usually attributed to the fact that all other things are *not* equal. Factors whose origins are external to the workplace are invoked to explain why the working class has failed to arrive at the elementary tenets of Marxism: for example, racism and the general ethnic heterogeneity of the U.S. workforce; the relative affluence of many U.S. workers (due in part to imperialism); the calculated manipulation of working-class consciousness by the schools, advertising, the mass media; and so forth.

Now there is no question that these "external" factors have played a major role in inhibiting the development of working-class consciousness in the United States. But Braverman's work suggests that the problem may not all be external to the work process; that even with "all other things being equal" certain objective features of the productive process in monopoly capitalism may militate against the development of proletarian class consciousness. We will argue that the "consciousness-raising" effects of the large socialized workplace are by no means as automatic as many Marxists have supposed. The work experience in monopoly capitalist society both *collectivizes* and *atomizes* the working class. Recognition of this contradiction leads us to propose the need for a sharp break from traditional approaches to organizing—particularly those being revived by many contemporary "Marxist-Leninist" groupings.

First, consider the subjective implications of what Braverman describes as "the degradation of work"—the historical process by which capital has appropriated to itself the scientific conceptualization and the technological mastery of the productive process. As Braverman puts it, the world of the worker becomes "increasingly devoid of any content of either skill or scientific knowledge"; the workers' "critical, intelligent, and conceptual faculties" are "deadened or diminished." In this situation it becomes impossible (or rhetorical) to raise the traditional revolutionary question: "What do we need the bosses for anyway?" The capitalist as *parasite*—as mere owner—is now disguised innocently as "management." It is the host of managers, engineers, supervisory personnel, planners, etc.—not the average workers—who now appear necessary and essential to the productive process (and, of course, *are* necessary, given the basic class antagonism inherent in capitalist production).

In this situation the prospect of workers' control—which is *the* socialist project in the workplace—appears faint indeed. As Braverman comments, unfortunately only in a footnote:

> The conception of a democracy in the workplace based simply upon the imposition of a formal structure of parliamentarism—election of directors, the making of production and other decisions by ballot, etc.—upon the existing organization of production is delusory. Without the return of requisite technical knowledge to the mass of workers and the reshaping of the organization of labor—without, in a word, a new and truly collective mode of production—balloting within factories and offices does not alter the fact that the workers remain as dependent as before upon "experts," and can only choose among them, or vote for alternatives presented by them. Thus genuine workers' control has as its prerequisite the demystifying of technology and the reorganization of the mode of production. (P. 445)

Second, let us take a closer look at the presumed "collectivizing" effect of the large socialized workplace. It cannot be denied that such a workplace brings large numbers of people together in a situation of shared oppression, but what is the quality of the social interactions which can develop among co-workers in a typical large-scale industrial setting? Does the productive process really generate a conscious sense of collectivity?

Again, leaving aside "external" factors such as racism and sexism (which obviously disfigure the kinds of social interactions which can occur at work or anywhere else), we would suggest that there are certain features of the productive process which militate *against* the development of a sense of collectivity. For example, one of the characteristic features of the capitalist workplace is the serial organization of production. In both the factory and the office it is rare for a group of workers to converge on the same material; instead they work in series, each separately adding some modification or part. When, in addition, serial production is spread out over vast physical expanses, the isolation of workers can be extreme. In a mile-long steel mill, workers in one department have no contact with those in other departments. A coke-oven worker and a rolling-mill crewman at opposite ends of a single plant might just as well be working at opposite ends of the country, for all that they are likely to come into contact.[1] Even on the much more densely populated auto assembly line:

> Social interaction tends to be restricted to superficialities, because the conditions of work on the assembly line make any sustained or deeper sociability impossible. On many jobs, noise makes it extremely difficult to be heard. In addition, a large number of jobs require close attention because of the speed of the line and the necessity to keep up. Restricted physical mobility also affects social interaction, limiting contacts to those directly at hand.[2]

Both the degradation of labor and what could be called the "de-collectivization" of labor are, to a certain extent, *built into the productive technology itself*. For example, to replace an auto assembly line with machinery conducive to creative and collective work would require a multi-billion-dollar capital investment. In such a case it can be said that bourgeois social relationships (exemplified by the hierarchical and atomized work process) are actually built into the existing machinery. Braverman approaches this conclusion, but quickly draws back from it with an assertion of the traditional dictum that it is not the technology, but only its present managers, which is at fault.[3]

On top of the technological factors which weigh against the formation of collective consciousness (and confidence) at the workplace, there are the more or less arbitrary, day-to-day interventions of management: harassment of individual workers who "step out of line" or "think they're smart"; rules against any kinds of innovations on the part of the workers, even when these do not interfere with the speed of the work (for example, General Motors' hostility to workers' "doubling up" to help each other);[4] explicit prohibitions against conversations among employees.[5] The deliberate efforts of management to make workers feel isolated and passive are particularly striking in the poorly mechanized human service settings. For example, in the hospital, where the nature of the work requires close interaction among workers and initiative on the part of even the "unskilled," there are often injunctions against fraternizing with workers of marginally different rank and penalties against workers who do seek to exercise initiative in the interests of good patient care.[6]

To the extent that such stratagems succeed (whether they are embedded in the technology or consciously pursued by management), to the extent that work has been degraded and de-collectivized, stripped of intellectual and social satisfactions —to that extent the basic contradiction of capitalism, the contradiction between the social nature of work and the private purposes of production, is muted, disguised, and for all practical purposes, liquidated. The "large socialized workplaces" which were supposed to generate the "gravediggers of capitalism" themselves become graveyards for human energy and aspirations. As Braverman makes so clear to us, they are *designed* to undermine the development of collective confidence and to reinforce feelings of powerlessness, anxiety, and social isolation.

The overwhelming response of working people is to withdraw their energy and hopes from the work world and invest them in the only other sphere provided by contemporary capitalism— *private life*. The counterpart to work life is not "the community" because the "community" is largely gone—bulldozed out of the way or replaced by self-enclosing tract homes. What is left—private life—consists of whatever nurturance can be extracted from the nuclear family; whatever sense of autonomy can be achieved through individual commodity consumption; and, in default of those sources of satisfaction, escapist fantasies, themselves commoditized in the form of mass-spectator sports, macho movies, and even the "news" (public events portrayed as the unfolding and intersection of public lives). Of course, a thousand and one factors external to the workplace, from advertising to car culture, promote the idea that private life is *real* life. But our point is that the privatization of working-class life is ultimately rooted in the workplace itself and in the capitalist degradation of work.

To sum up where we have gotten to: our brief pursuit of "the subjective aspects of the development of the working class" has led us along a trail which begins in the degraded (and often isolated) experience of work and ends in the isolated (and often degraded) experience of private life. This argument stands *in contradiction* to the traditional Marxian wisdom about the development of working-class consciousness. If (as tradition holds) the conditions of work in monopoly capitalism can lead to revolutionary proletarian consciousness, we would suggest they can also lead to the dispersion of collective consciousness in *any* form. If they can lead to the consolidation of the class, they can also lead to its *atomization*. And in the first seven decades of this century, there can be no question that it is the latter part—the negative part of the dialectic—which is ascendant.

We want to conclude with two brief observations on strategy for the socialist movement:

(1) It seems to us that there is no longer any reason, other than a romantic one, to insist on the unique centrality of the workplace as the locale for the development of class consciousness in the United States. Parallel to the withdrawal of working-class energies from the workplace there has been a displacement of the contradictions engendered by the capitalist mode of production to other spheres—the family, the schools, even the health system. The schools attempt to bear the burden of the reproduction of work force stratification; the health system expands to absorb the wounded, or at least, the wounds inflicted at work, from lung damage to neurosis. The family (once a center of production itself) struggles along under the weight of all the psychological needs which can't be met at work or in an outside social life. Out of these "displaced" contradictions have grown movements which could not have been predicted by an analysis focused narrowly on the point of production: for example, the women's movement, the student

movement, movements of welfare and health care recipients. At the same time, institutions other than the workplace—urban neighborhoods, prisons, college campuses, etc.—have emerged as major sources of collective consciousness and militant struggle.

A question may be raised as to the relevance of such "special interest group" struggles to the development of overall class consciousness and class struggle. But, as Lenin argued long ago in *What Is To Be Done?*, the same question may also be asked of workplace organizing:

> The economic stuggle merely brings the workers "up against" questions concerning the attitude of the government towards the working class. Consequently, *however much we may try* to "give to the economic struggle itself a political character" *we shall never be able* to develop the political consciousness of the workers (to the degree of Social-Democratic consciousness) by confining ourselves to the economic struggle, for *the limits of this task are too narrow.* . . .
>
> The workers can acquire class political consciousness *only from without,* that is, only outside of the economic struggle, outside of the sphere of relations between workers and employers. The sphere from which alone it is possible to obtain this knowledge is the sphere of relationships between *all* classes and the state and the government—the sphere of the inter-relations between *all* classes.

In the workplace, as anywhere else, only conscious political efforts can bridge the gap between the immediate self-interest struggle of a particular group and the larger aims of a socialist movement.

(2) However, the workplace does remain a central arena for organizing. There is no question that the work experience is a major determinant of people's consciousness. But exactly how it shapes people's consciousness is not something that can be predicted with simple formulas about the effect of the "large socialized workplace." We need, obviously enough, to start with what is, and not with what should be (or should have been). *What is* may often be discouraging, but there can be no escaping into an analysis so abstracted from reality that it has become, in essence, idealist.

The best starting point, it seems to us, is to accept the most pessimistic implications of Braverman's analysis: modern industry was designed to make workplace struggle not only difficult, but *unimaginable*. Realizing this, we understand that the really amazing thing is, that despite all the Taylorization and fragmentation and degradation of work, despite all the things that make people feel stupid and worthless and isolated, on the job and off, people do manage to survive. And not only to survive but to hold on to some sense of dignity and determination. Even in the most highly "rationalized" industries, people (workers) find ways to subvert the technology and the supervisors, to hold down the speed of work, to gain a little breathing space, to build informal networks of resistance and support.[7] Many of the responses to the workplace are individualistic and escapist (dope, absenteeism, etc.—though even these must be seen as valid human responses to workplace oppression), but others contain the germs of collective power. It is here, we would argue, that organizing has to begin: with a conscious effort to learn and build on all the mechanisms by which people *do* survive, resist, and fight back.

Notes

1. An example of the significance of this lack of contact: In steel mills, typically coke-oven and blast-furnace workers are black, while rolling-mill workers are white. The black workers are paid worse, suffer worse working conditions, have fewer opportunities for advancement, and are harassed more by the foremen. White workers in other departments, having no contact with the coke-oven and blast-furnace workers, and having their own grievances (though often of somewhat less severity) find it hard to sympathize with the claims of blacks that the blacks are being singled out for worse treatment. They often consider black demands as more or less those of "cry babies" demanding special privileges. The problem is both exacerbated by and itself exacerbates racism.
2. Robert Blauner, *Alienation and Freedom* (Chicago, 1964), p.

114. We are not arguing that the "de-collectivizing" impact of the modern workplace outweighs its "collectivizing" impact: the two thrusts are dialectically related. This dialectic continues when we look at unions, whose existence is usually taken as proof of the collectivizing impact of the workplace. Unions are the institutional expression of the workers' collective power, but at the same time unions function to *limit* the workers' sense of collective power: in a day-to-day sense, the unions suppress workers' militancy as it arises spontaneously on the shop floor. In a long-term sense, the union limits the content of collective struggle to those issues which the union defines (for legal, traditional, or bureaucratic reasons) as contractually relevant: wages and hours are relevant; the organization of work, the technology employed, the nature of the product are irrelevant. Thus both direct activity by the workers themselves and the issues which are most central to the work experience as a *common* experience are defined as irrelevant by the only legal form the workers have. The unions, far from becoming "schools for communism," become (in John Welch's phrase) the "institutional form of the immobilization of the working class."

3. *Labor and Monopoly Capital,* pp. 193-95 and pp. 227-33.
4. Stanley Aronowitz, *False Promises* (New York, 1973), p. 23.
5. Some years ago, one of the authors worked on one of several machines that printed address labels for the *Saturday Evening Post*. Although the machines were almost entirely automatic, the machine operators were required to stand beside their individual machine at all times. This made conversation impossible because of the noise of the machines. The rule was justified on the grounds that "it would look bad" if someone from higher management came down and saw the workers standing around talking.
6. See John and Barbara Ehrenreich, "Hospital Workers: A Case Study in the New Working Class," *Monthly Review,* January 1973.
7. See, for example, M. Guttman, "Primary Work Groups," *Radical America,* Vol. 6, no. 3, pp. 78-91.

Capitalist Efficiency and Socialist Efficiency
by David M. Gordon

Marxism has been blooming around the world. Paul Sweezy and Harry Magdoff call it "a veritable renaissance."[1] Samir Amin celebrates an "extraordinary rebirth."[2] Just as socialist practice has deepened and spread its roots internationally, so has Marxian theory borne increasingly rich and bounteous fruit.

The Marxian renaissance has involved some theoretical tussles with the earlier Marxist orthodoxy. These theoretical reformulations have necessarily focused on the central problematic of the materialist perspective—on its analysis of the *production process* itself. Many Marxian theorists have contributed to our revitalized understanding of production.[3] Through their work, we can more clearly appreciate several important dialectical relationships: between the mode of production and social formations, between the forces and relations of production, between technology and job structure, between technical imperatives and class struggle. With *Labor and Monopoly Capital*, Harry Braverman has provided some central elements for this clarified focus, helping arm us with many critical arguments for our assault on traditional views.

But we have only begun to develop this new theoretical view fully and rigorously. The recent work on production poses some central theoretical questions which have not yet, in my view, been fully appreciated. In this short essay, I want to focus on three such questions, all interconnected, all fundamental to the further development of this theory. These three questions can be simply (if ambitiously) formulated in preface:

(1) Is it possible for capitalists to deploy technologies and job structures which "control" workers if those elements of production are not "cost-minimizing"?

(2) In a period of socialist transition, why is it not in the

interests of workers to develop the most "efficient" production process, expanding the forces of production as rapidly as possible, even though such a production process might "degrade" them to a "fragment of their former being"?

(3) If capitalist production relations affect workers' class consciousness, do they provide the basis for a revolutionary force not only to overthrow capitalist relations but also to sustain continual movement and struggle along the socialist path?

All three questions float with grandiosity on the highest clouds of theoretical abstraction. At best, in such a short essay, they can barely be posed with clarity, much less answered with precision. In developing their central importance here, I want simply to argue that each requires a new and more rigorous formulation of the two central concepts of my title: *capitalist efficiency* and *socialist efficiency*.[4]

Capitalism and Control

Braverman and Marglin most clearly focus the thrust of the new approach. The simplest possible summary of their argument follows these contours: There are no universally efficient forces and relations of production for all societies. Capitalism has developed a production process which not only delivers the goods but also controls its workers. We need not assume that socialist production must adopt the same kinds of technology and job structures in order to achieve (at least) the same level of living standards. Other relations of production— less degrading and routinized—might conceivably match or outstrip the productiveness of capitalist relations.

This argument has had wide appeal, apparently for two main reasons. First, it has spoken directly to the widespread concern with "alienation" and "job dissatisfaction" which seemed to surface in many countries in the 1950s and 1960s. Second, it provided a theoretical justification for the growing Marxian critique of production in the Soviet Union, where it appears that workers confront job cells with specialization and hierarchy rivaling those in advanced capitalist formations.

Popularity does not guarantee clarity. I think that some imprecision has been clouding the importance of this new approach. Whether from orthodox or Marxian social scientists, many of the same questions seem persistently to recur.[5] They typically reveal a common logical sequence:

Q: What manifests the "werewolf hunger" for capital accumulation? A: Competition. Q: If competition reduces capitalists to reflexive accumulators, will it not be necessary for every capitalist to adopt the same production process? A: Yes, at the abstract level, because competition is the great equalizer. Q: Won't it also turn out that those manifest forces and relations of production apply the "most efficient" technology and job structure? A: Yes, since those who did not employ the "most efficient" production process would lose in the competitive battle to those who did. Q: In that case, isn't it sufficient to speak of "efficient" technologies? What can we possibly add by speaking of technologies (or job structures) which "control" workers? A: Some technologies are better at controlling workers than others. Q: How can there be some technologies which differ from others? Isn't it true that there will be only one kind of technology adopted—the most "efficient" kind? How can there be some "more controlling" technology which is not "most efficient"? A: What you call most "efficient," I call "'controlling." Efficiency is irrelevant by itself if the capitalist cannot produce surplus value. Q: But if there is only one kind of technology which triumphs under capitalism, it must be the most efficient technology possible—whether or not it controls workers. Will it not therefore be true that all other technologies —potential or actual—will be less efficient at a given stage of development? A: Some other technologies may be less controlling. Q: Isn't efficiency all that matters? A: Control matters more. Q: Efficiency! A: Control!

Is this simply an impasse? Is the analysis of efficiency and control like comparing apples and oranges? Are the concepts incommensurable?

Since I have heard these kinds of questions many times recently, I believe that they represent a significant response to the recent literature on production. I do not think that they have been adequately addressed.[6] They deserve careful response, for they raise some central questions, as we shall see, about the transition from capitalism to socialism.

I have found it most useful to reformulate these questions within the following framework:[7]

(1) In any class society, a mode of production can continue to dominate if and only if prevalent production processes reproduce the class relations defined by (the logic of) that mode of production. This requires a growth in the forces of production which is consistent with a particular pattern of class domination. It requires a set of social relations of production which reproduce ruling class power.[8]

(2) In a specific class society, therefore, one cannot speak of the forces of production "in general."[9] The forces of production are manifested in a historically specific production process. That production process either tends to reproduce ruling class dominance or tends to erode it.[10] The same production process might have completely different effects in two different social formations.

(3) The "efficiency" of a production process, therefore, can be considered conceptually in two ways: efficiency has both a *quantitative* and a *qualitative* aspect.

In general, a production process is *quantitatively* (most) efficient if it effects the greatest possible useful physical output from a given set of physical inputs (or if it generates a given physical output with the fewest possible inputs).[11] I can think of no theoretical reason why there would not be many (if not an infinite number of) possible production processes with equivalent quantitative efficiencies at any stage in the natural development of the means of production—in physical terms—in any given society.[12]

In class societies, a production process is *qualitatively* efficient if it best reproduces the class relations of a mode of production.[13] In more specific terms, a production process is qualitatively (most) efficient if it maximizes the ability of the ruling class to reproduce its domination of the social process of production and minimizes producers' resistance to ruling class domination of the production process.[14] Given the opposition between the ruling class and direct producers, it would be surprising if production processes in a social formation *stably* dominated by a mode of production did not tend toward the most qualitatively efficient forms possible.[15]

With these three simple theoretical axioms, it seems possible to recast our archetypal Q & A exchange in more fruitful terms. Three frames of reference prove helpful—theoretical, historical, and comparative static.

Theoretically, we can begin to avoid the trap into which some Marxists, it appears, are typically prone to fall. I often hear debates about the "driving forces" of capitalism. What keeps the system on the move—*competition* or *class struggle*? Some of those debates seem to presume that those two answers are mutually exclusive. In the terms proposed above, however, they are clearly interdependent. The dialectics mediating the relation between the imperatives of quantitative and qualitative efficiency also govern, in turn, the relation between the forces of competition and the forces of class struggle. Competition enforces capitalist concern with quantitative efficiency. Class struggle enforces capitalist concern with qualitative efficiency. The unity of production kneads together the two imperatives.

Historically, with the development of industrial capitalism, capitalists searched for any (among the many possible) quantitatively efficient production processes. Growing working-class resistance forced them to search for the most qualitatively efficient production processes. The search began on an exploratory basis. Those capitalists who discovered the most successful combinations gained comparative advantage over their competitors—not necessarily because their costs were minimized, in some prices-of-production sense, but because they were better able to discipline their workers, avoid strikes, and extract surplus product from their labor. These qualitatively efficient processes were copied, over time, and became more and more prevalent. Other quantitatively efficient processes fell away because they were less qualitatively efficient than those which triumphed.[16] As capital accumulation continues, this process of exploration, discovery, and diffusion itself proceeds apace. As Braverman shows, its pace probably intensifies.

There are two different ways to formulate this same kind of argument about capitalist production in *comparative static* terms—a weaker and stronger version.

More weakly, we can propose a simple formulation. At any moment in capitalist societies, given relative factor prices, there exists a set of quantitatively (most) efficient combinations of productive factors with more than one (and probably many) constituent elements.[17] Within that possibility set, capitalism will tend to develop those production processes which are qualitatively most efficient. In more orthodox language, this formulation suggests that capitalist production processes maximize *qualitative* efficiency subject to the constraint that they are *quantitatively* efficient.

There is a stronger version toward which, I gather, some Marxists would lean. This version suggests that the criteria of quantitative and qualitative efficiency tend somewhat to conflict in capitalist societies. The requirements of labor control tend to push capitalists, according to this version, toward production processes which are less quantitatively efficient than other (technically) possible processes.[18] This version suggests that, at least from time to time, the set of qualitatively (most) efficient production processes does not intersect with the set of quantitatively (most) efficient production processes. Capitalists are forced, in other words, to accept sacrifices in potential physical output, given prevalent scarcities and relative factor prices, in order to maintain worker discipline and reproduce their control over the means of production.[19] In more orthodox language, in this case, capitalist production processes maximize *quantitative* efficiency subject to the constraint that they are *qualitatively* efficient.

This stronger version would help explain, for many, the apparent anomalies revealed in the literature on job design experiments. Virtually without exception, that literature suggests that worker productivity would increase if jobs provided more varied, less "degraded" tasks. And yet capitalists do not implement those kinds of job changes. One might hypothesize, in the theoretical language proposed here, that capitalists are unable to make quantitatively efficient changes in job structure because those changes would involve (relative) qualitative inefficiencies.[20]

These theoretical, historical, and comparative static formulations can help us, I think, clarify the meaning of some classic Marxian hypotheses about accumulation tendencies under capitalism.[21] Two related examples can suggest their general usefulness.

First, Marxists have traditionally hypothesized the continual *concentration* of capital under capitalism. Why? It seems extremely unlikely, as some orthodox economists might maintain, that the progressive concentration of capital always generates (monotonically) more quantitatively efficient production processes. Indeed, we have many reasons for thinking that technical "economies of scale" do not require anything like the concentration of capital developed in the advanced capitalist stages of accumulation. For a wide variety of reasons, however, it does make sense that the continuing concentration of capital is required by the imperatives of qualitative efficiency under capitalism; capitalists may be continually driven, other things equal, to increase the mass of "dead labor" confronting workers in order to minimize workers' potential resistance to their exploitation.[22]

Second, Marxists have long debated the "law of the tendency of the rate of profit to fall." At the theoretical level, is that tendency logically necessary? That question can be decomposed into several parts.[23] One of its constituent questions focuses on the necessity for the technical composition of capital to rise. Why should capitalists always replace workers by machines? In terms of quantitative efficiency, indeed, it may turn out that a rapid expansion of the reserve army of labor may, from time to time, produce a sufficiently large drop in the labor market wage that capitalists are induced to invest in "labor-using" rather than "labor-displacing" production processes.[24] It makes sense to argue, however, that imperatives of *qualitative* efficiency continually force capitalists to substitute machines for workers in pursuit of greater control over production and that, over the long run, these imperatives will be sufficiently strong to overcome any countervailing changes in relative factor prices. In this sense, at least, the notion of *qualitative* efficiency can make the "necessity" of a continual rise in the technical composition of capital more theoretically plausible.[25]

There are many problems with this kind of theoretical formulation. At best I think it serves to clarify some questions for further discussion and research. In this stage of my argument, two simple implications seem most important to summarize.

First, I think that this formulation can help focus many

of our continuing analyses of tendencies under capitalism—on the causes and consequences of crisis, for instance, or the continuing centralization of capital, or the progressive internationalization of capital. For those analyses, we can apply the concept of *capitalist efficiency* in a very precise sense: production processes embody capitalist efficiency if they best reproduce capitalist control over the production process and minimize proletarian resistance to that control. As capitalism develops and as workers continually develop their organized capacity to resist capitalist exploitation, it seems logical that these imperatives of qualitative efficiency would become increasingly determining.

Second, I think that this way of talking about efficiency can be critical in our attempts to pose the theoretical and ideological contrasts between capitalism and socialism.[26] If capitalist production relations embody capitalist efficiency, it makes common sense that socialism would permit the development of productive forces with equivalent quantitative efficiency which would be much more liberating of workers' potential. Can we be more precise? This requires a sharp left turn toward the discussion of socialism itself.

The Socialist Path

The theoretical analysis of socialism has been evolving rapidly during the current period of renaissance. I need to draw on that theoretical discussion to develop the themes of this essay somewhat further.

Most Marxists agree that socialism is a transitional (rather than a permanent) formation. Societies progress along the path of socialist transition, from this perspective, if they move farther and farther away from the domination of the capitalist mode of production and move progressively closer to the ultimate goal of "perfect communism."[27] One speaks, therefore, of a "theory of socialist transition" rather than a "theory of socialism."

Recent debates about the theory of socialist transition have largely focused on the different experiences of the Soviet Union and China. Are they both still socialist? Has the Soviet

Union become a "state capitalist" formation? Has China been backsliding down that revisionist road after the halcyon years of the Cultural Revolution?[28]

Different theoretical premises are buried in those debates. Partly because other issues intrude, partly because the theory of socialist transition has always remained imprecise, those theoretical issues are rarely posed clearly.

I think that a reformulated concept of *socialist efficiency*, building on the distinction between quantitative and qualitative efficiency, can help clarify the appropriate theoretical questions—and, with that clarification, can help sharpen the debate.

I would suggest that two main logical conceptions have dominated traditional Marxian views of socialist transition, however imprecise those views have seemed. Each of those conceptions corresponds to a different vision of the ultimate communist objective.[29]

The first view is simpler. I shall call it the "free labor" view of socialism and communism. It projects a vision of perfect communism as a classless society in which all people are free of nature's "whip of hunger." Necessary labor time is reduced to the barest minimum, freeing everyone to realize their human potential as creative people. Nothing more is specified about the prospective social relations of this perfect communist state. Communist society will have only two determining characteristics: (1) insignificant necessary labor time requirements (2) equally divided among all members of the society.

In its simplest possible version, this specification of the communist objective suggests an elementary view of socialist transition.[30] Two conditions are necessary and sufficient to guarantee continual movement along the path toward communism. First, the productive forces must develop continually until they reach the point where necessary labor requirements become insignificant. Second, workers must maintain sufficient control over the means of production so that this continual reduction in necessary labor time requirements is equally distributed among all members of the society.[31]

The second view of socialist transition adds to the first. I shall call it the "collective labor" view of socialism and communism. It projects a vision of perfect communism as a classless society in which people *both* enjoy freedom from the "whip of hunger" *and also* share in mutual, respectful, reciprocating, collective social relationships. Necessary labor time is reduced to the minimum, of course, but another specification is added. Communist society, in this view, will have three determining characteristics: (1) insignificant necessary labor time requirements (2) permitting shared collective social relationships (3) through which everyone is equally free *and* responsible.

This second specification of the communist objective involves a somewhat more complicated view of socialist transition.[32] Three conditions are now necessary and sufficient to guarantee continual movement along the socialist path. First, the productive forces must develop continually, just as before. Second, the relations of production must evolve in a direction which continually develops workers' capacity to share equally in mutual, responsible, and collective social relationships.[33] Third, workers must maintain sufficient control over the production process so that both this reduction in necessary labor time and this developing collective capacity are equally distributed among all members of the society.

To the extent that these two conceptions differ logically, I would argue that their differences are profound. In order to keep the discussion brief, I shall refract those differences through the prism provided by the earlier discussion of efficiency.

Since both views of socialist transition build upon the continual expansion of the forces of production, both are concerned with quantitative efficiency. At any moment along the transitional path, other things equal, a socialist society will prefer one production process over another if the former is quantitatively more efficient.[34]

What about qualitative efficiency? It was defined earlier with respect to class societies. For the purposes of the theory of socialist transition, however, that earlier concept of qualitative efficiency has a clear "dual" or analog. In class societies, as defined above, a production process is qualitatively efficient if it best reproduces the class relations of a mode of production. Along the path of socialist transition, in reverse, a production process embodies socialist (qualitative) efficiency *if it best supports movement along the path toward a classless society. A production process embodies socialist (qualitative) efficiency, more specifically, if it maximizes the ability of the working class to increase its domination of the means of production and minimizes the possibility of revisionist slippage back toward further ruling class domination.*[35]

In these terms, the two views of socialist transition obviously suggest different definitions of the necessary elements of qualitatively efficient production processes in socialist transition. The "free labor" view involves a loosely specified notion of socialist qualitative efficiency. Any production process which reinforces working-class control of the means of production satisfies the qualitative imperative from this perspective. At the extreme, a production process which "degraded" workers to a "fragment of their former being" would *not* be inconsistent with this definition of socialist qualitative efficiency as long as that process did not interfere with the continuing control of the means of production by workers or their designated representatives.[36]

The "collective labor" view more narrowly specifies the criterion of socialist qualitative efficiency. Along the path of socialist transition, from this perspective, production processes will achieve socialist qualitative efficiency if (and only if) they best support the development of producers' capacities to share responsibly and collectively in the relationships of their daily lives. Work processes which degrade workers' skills would be inconsistent with this criterion precisely because they would reduce, rather than augment, workers' capacities to control and coordinate the social relations of production. Lenin wrote that the "withering away of the state" becomes possible when "*all* have learned to administer and actually to independently administer social production. . . ."[37] In the same way, and at a somewhat more fundamental level, we might write that *the "withering away" of class control of production becomes possible only when all have learned to administer social production.* Production processes would satisfy the criterion of socialist qualitative efficiency if they best supported this development.

The difference between these different conceptions of socialist efficiency is fundamental. Debate cannot be waged by shotgun fusillades of selective quotations from the classic texts, for I think that both positions are adequately represented in the works of Marx, Engels, Lenin, and Mao.[38] The debate must rather be conducted through a focused and rigorous discussion of the theoretical, political, and ideological implications of each view.

Although I have tried to formulate the two perspectives fairly and judiciously, I have strong personal preferences for the "collective labor" view of socialist transition.

I would argue, on the one side, that the "free labor" view accepts several elements of the bourgeois perspective as premises for its position. It views history solely in terms of peoples' struggles to master nature, ignoring the necessarily interdependent character of the relations *among* people in society.[39] It projects a largely individualist view of perfect communist society, in which people would be free "to do their own things" as individuals without specified regard for the impact of "their own things" on others. It tends to define socialist transition almost wholly in terms of the forces of production, levels of relative consumption, and mechanisms of distribution, largely ignoring the form and content of the social process of production and its impact on peoples' own development as people. For those reasons, I think that it suffers serious problems of underspecification. In particular, it does not articulate a theoretical basis for the solution to two obvious problems of socialist transition. First, how will people develop the *capacities* to realize their "creative potential" if those capacities are not, in fact, directly developed through peoples' experiences as producers along the transitional path? Second, how will people be able to guard against the resurgent power of some "old" or "new" ruling class if they do not directly develop the capacity to administer their own affairs through their evolving production experiences?

I accept that there are many problems with the "collective labor" view as well. I think that it is more consistent with a "proletarian" view of peoples' interdependence as social producers, but there is too little space here to try to develop the logical basis for that kind of argument.[40] At the least, in my view, that perspective helps focus discussion of socialist transition in countries which have already experienced "socialist" revolutions: it helps us evaluate their experiences not in terms of which country pursues more "correct" politics, by some absolute measure, but rather in terms of how the developing production processes of each country appear through the lenses of socialist qualitative efficiency criteria from the "collective labor" perspective.

And this discussion of socialist transition also helps us return to the ultimate destination of this lofty journey—to some critical questions about the relationship between capitalist accumulation and the revolutionary socialist movement.

Capitalism and Socialist Revolution

For Marxists, these discussions of capitalism and socialism are not disjointed. Marxists have traditionally argued that the contradictions of capitalist development would internally generate the seeds of socialist revolution. If we assume that workers both carry out anti-capitalist revolution and sustain the socialist struggle after that revolution, then the links between capitalist and socialist "efficiency" are automatically forged by our theoretical concerns.

What are those links?

Marxists have traditionally identified two major aspects of capitalist development which contain potentially revolutionary contradictions.[41] These two aspects are logically linked, I think, to the two prevailing views of socialist transition.

The first view focuses on the irrationality of capitalist distribution. I shall call this the "unnecessary labor" view of capitalist contradictions. It suggests that the development of capitalism witnesses an unfolding contradiction between the progressive expansion of the productive forces—which would *potentially* permit reductions in necessary labor time—and the concentration and centralization of capitalist control—which *prevents* workers from realizing those potential reductions in necessary labor time. Workers are more and more likely, according to this view, to recognize their common class interests in revolting against a capitalist class which lives off their labor and precludes their taking advantage of the labor-reducing potential of the productive forces. They will revolt, in short, to convert "unnecessary labor" into "free labor."

The second view focuses on the contradictions of capitalist production itself. I shall call this the "atomized labor" view of capitalist contradiction. It suggests that capitalist development witnesses an unfolding contradiction between the increasing "objective interdependence" among producers through the developing division of labor—which dramatizes the *collective character* of social labor—and the atomization of both labor power and labor activity—which reduces workers to fragments of their former selves and pits them in competition among themselves. More and more, in this view, workers are bound to recognize their objectively common interests as socially collective producers in revolting against a production system which precludes their developing and realizing that collective force, which limits them to a pale, powerless image of their potential class power. They will revolt, in short, to convert "atomized labor" into "collective labor."

There are many contexts in which these two views of capitalist contradictions are not necessarily opposed. Indeed, they were linked for Marx and Engels in the nineteenth century. The development of the nineteenth-century capitalist factory appeared to provide the basis for the "fusion" of those two contradictions.[42] The factory joined together many workers who, through the simple presence of the boss, could clearly see the irrational allocation of the fruits of their "unnecessary labor." The factory also forged those workers together into an objectively powerful productive collective, welded together by "modern machinery," progressively recognizing their interdependent power as they became "operatives" of that social capital. Who needed to estimate the relative importance of those two sources of rebellion when they were fusing through a unifying class?

Advanced capitalist accumulation has deflected those two contradictions. The "unnecessary labor" contradiction has been complicated by the growth of bureaucratic hierarchy—confusing the visible patterns of surplus labor consumption—and the global spread of capital—providing some margin for the progressive improvement of workers' real living conditions in the advanced capitalist countries. The "atomized labor" contradiction has been short-circuited in two ways: First, as Braverman so dramatically demonstrates, workers no longer "operate" those machines but now simply "observe" them; with that change, workers' sense of their potential collective power doubtlessly diminishes. Second, advanced capitalist factories tend more and more to separate workers both within and between factories, "stretching out" stations within the shop and decentralizing plants across the extended capitalist terrain.[43]

Where to struggle? How to win?

To the extent that the revolutionary potential of these two contradictions has been muted, socialists must struggle to lay them bare once again. In debates about revolutionary strategy, Marxists have too rarely linked analyses of capitalist development to theories of socialist transition. Intervention through anti-capitalist struggle must be strategically predicated on some clear theoretical understanding of the *kind* of socialist movement a revolutionary working class would potentially sustain. Working-class rebellion cannot simply be *against* capitalism. It must also struggle *for* socialism.

What kind of socialism? The choice between the "free labor" and "collective labor" views of socialist transition becomes critical. While there is too little space here to explore the kinds of divergent organizing strategies which might derive from those two different views, I believe that such analyses are urgently required. It should be obvious, as a preface to those discussions, that I believe that the "unnecessary labor" view of capitalist contradictions provides an insufficient basis for revolutionary socialist strategy. Although the classical nineteenth-century framework for the "atomized labor" view of contradictions must be revised, I think that revolutionary socialist strategy needs to pay more and more attention to that strand of capitalist development. Workers have been atomized in new ways under advanced capitalism, but they still experience that atomization objectively and, no doubt, have subjective responses to it.

And here, we can finally arrive at some concluding comments about Harry Braverman's book. His contribution flows through his exposure of the continuing atomization of work under advanced capitalism. But the book's weakness flows from that same strength. Because Braverman explicitly avoids discussing the impact of labor in monopoly capitalism on workers' consciousness—and he has every right to avoid that discussion—he leaves us open to a misimpression. Without any warning to the contrary, Braverman's story of the continuing degradation of labor in the twentieth century might lead us to the conclusion that the "atomized labor" contradiction will soon erupt, however belatedly, in exactly the same form—through the capitalist "factory"—as Marx and Engels projected for the nineteenth century. In fact, the content of Braverman's tale of capitalists' progressive scientific management of labor might lead to exactly the opposite conclusion: that workers' potential revolutionary realization of their collective potential as interdependent producers will not come through their lives in the factory, where their daily experiences increasingly belie that prospect of potential power, but through some other aspects of their daily experiences under capitalism.

Which aspect? How will it happen? That is the fundamental question to which both Braverman's book and this essay have led us. *Capitalist efficiency* has changed the terms of workers' collective experience as workers. The imperatives of *socialist efficiency*, seen through the "collective labor" lens, require our intensifying focus on that collective experience. Through that focus, we can hope, a clearer sense of revolutionary socialist strategy will emerge.

Notes

1. "Twenty-Five Eventful Years," *Monthly Review*, June 1974, p. 4.
2. Untitled Essay, *Monthly Review*, June 1974, p. 19.
3. Among many others, those works which have contributed to my own understanding include Harry Braverman, *Labor and Monopoly Capital* (New York, 1974); L. Althusser and E. Balibar, *Reading Capital* (New York, 1968); Stephen Marglin, "What Do Bosses Do?" *Review of Radical Political Economics*, Summer 1974; Samir Amin, "Mode of Production

and Social Formation," *UFHAMU*, 1973; Frank Roosevelt, "Cambridge Economics as Commodity Fetishism," *Review of Radical Political Economics*, Winter 1975; Catherine Stone, "The Origin of Job Structures in the Steel Industry," *Review of Radical Political Economics*, Summer 1974; Samuel Bowles and Herbert Gintis, *Schooling in Capitalist America* (New York, 1976); Herbert Gintis, "The Nature of the Labor Exchange: Toward a Radical Theory of the Firm," Harvard Institute of Economic Research, mimeo, 1973; Stanley Aronowitz, "Marx, Technology, and Labor," mimeo, 1975; Richard C. Edwards, "Alienation and Inequality: Capitalist Relations of Production in Bureaucratic Enterprises," Ph.D. dissertation, Harvard, 1972; and James O'Connor, "Productive and Unproductive Labor," *Politics and Society*, Vol. 2. I have tried to contribute to this process through David M. Gordon, *Theories of Poverty and Underemployment* (Lexington, Mass., 1972), especially Chapter 5; and David M. Gordon, Richard C. Edwards, and Michael Reich, "Labor Market Segmentation in American Capitalism," mimeo, 1976.

4. The discussion in this paper will be brief and elementary for reasons of space and preparation. I am trying to develop several themes from this paper in more detail. I would welcome responses and suggestions from readers toward that end.

5. The responses are often similar. In this context, that should not be so surprising. There is a strong streak of technological determinism in both the bourgeois and the orthodox Marxist traditions.

6. These questions are not directly considered in Braverman's book and his response to them is not clear to me. One who acknowledges these questions and begins to respond to them is Stephen Marglin, "What Do Bosses Do?: A Postscript," Harvard Institute of Economic Research, mimeo, 1975.

7. I apologize for some of the jargon here. If ponderous, jargon also sometimes permits brevity. For a full development of the sense in which I use many terms here, see Amin, "Mode of Production and Social Formation."

8. For further elaboration of these terms, see E. Balibar, "The Basic Concepts of Historical Materialism," in Althusser and Balibar, *Reading Capital*.

9. This point simply echoes Marx's many methodological warnings about the specificity of individual modes of production and the dangers of arguing about society "in general."

10. Could there be some equilibrium, a balance of forces? Yes, theoretically, but the presumed intensity of the class struggle in any social formation makes it unlikely that such an "equilibrium" would be very stable.
11. In capitalist societies, where inputs have prices, this formulation in physical terms is analogous to the conventional neoclassical formulation in money terms.
12. This could be formulated in mathematical terms, focusing on the underdetermination of the combinations of factors in a productive process. Some of this argument bears resemblance to the Cambridge (England) side of the "Cambridge Controversies" in economics.
13. What constitutes "qualitative efficiency" from the ruling class perspective, of course, represents exploitation from the perspective of the direct producers.
14. This distinction between quantitative and qualitative efficiency is not, I believe, formulated in Marx in exactly this way. The sense of the distinction runs parallel to Marx's discussion of the quantitative and qualitative aspects of commodities, on one side, and the "formal" and "real" submission of labor to capital, on another side (see Marx, *Capital,* Vol. 1). After formulating this argument in these terms, I discovered an almost exactly similar formulation in Aronowitz, "Marx, Technology, and Labor." That paper helped me clarify some of the arguments developed here.
15. This definition of the *qualitative* aspect of efficiency embodies what Braverman, Marglin, and others call "control." I find it useful to recast the definition in these terms in order to emphasize the *connections* between these two aspects of the production process and to help overcome some of the apparent "apples-and-oranges" incommensurability of "efficiency" and "control" in the recent literature.
16. This language avoids, I hope, any suggestion of "conspiracy," of capitalists plotting in the back rooms and planning, at a single stroke, these qualitatively efficient processes. I have tried to provide some interesting historical examples of this historical path in an essay in progress, "Toward a Critique of CAPITALopolis: Capitalism and Urban Development in the United States."
17. Expressed in slightly different terms: Given the existing development of the productive forces, there are many (technically and socially) possible concrete applications of those

productive forces which would take equally full advantage of the productive potential latent in that development.
18. By implication at least, these production processes are technically possible under capitalism but are logically inconsistent with the capitalist mode of production.
19. Some historical examples of this stronger case are summarized in Gordon, Edwards, and Reich, "Labor Market Segmentation in American Capitalism."
20. See Gintis, "The Nature of the Labor Exchange," for further development of this point.
21. It is my impression that either of the two comparative static versions is consistent with the presentation by Braverman in *Labor and Monopoly Capital*.
22. There is no space to develop this point here. One obvious example which supports that view is the rise of labor unions and their growing power through the process of capitalist development. For some further discussion of some of these issues, see Maarten de Kadt, "Management in Monopoly Capital: The Problem of the Control of Workers in Large Corporations," Ph.D. dissertation, New School for Social Research, 1976.
23. For instance, it involves separate questions about the technical composition of capital, the transformation between the technical composition and the organic composition of capital, and the relationship between the limits to the rate of profit and the actual rate of profit.
24. This possibility is clearly formulated by Maurice Dobb in the chapter on "Crises," in *Political Economy and Capitalism* (London, 1937).
25. While this formulation has not been further developed in writing, I and several colleagues at the New School for Social Research are working on this clarification.
26. Indeed some concepts like these are necessary in order to permit comparison across social formations dominated by different modes of production.
27. These basic points are made, of course, by Marx in the *Critique of the Gotha Programme* and by Lenin in *State and Revolution*.
28. For some of this kind of discussion, for instance, see the special issue of *Monthly Review* on "China's Economic Strategy," July-August 1975.
29. Some of these points are similar to those made by Howard

Wachtel, "Some Reflections on the Theory of Socialist Transition," to appear in a volume of essays in honor of Paul Sweezy, although we express the points in different ways.

30. There has been a recent formalization of this view, for instance, in a review essay on Marx by J.E. Seigel, *New York Review of Books,* October 31, 1974.

31. There are obviously some problems with this "simplest" logical formulation which require a somewhat more complicated view. In particular, there are potential interactions between the increase in the productive forces and the distribution of its products which require, operationally, further specification of the model. But the basic points remain the same.

32. Although I think this view is clearly implicit in much Marxian discussion of socialist transition, I do not, in fact, know of any modern sources where it is stated in so many words.

33. This is my version of what Marx meant when he talked about "winning the battle for democracy." This is also my version of what Lenin meant, in *State and Revolution,* when he talked about workers developing the capacity for "self-administration."

34. Since quantitative efficiency was defined in physical terms, the concept applies to a socialist society even if it does not rely on market prices as allocative guides. Pricing and allocation questions, if they arise, become second-order problems.

35. I use the term "ruling class" here, rather than "capitalist class," in order to keep open the analytic possibility that the "ruling class" emergent after a socialist revolution might be some "new class," like a "state capitalist" class.

36. I cannot emphasize how important this logical point is for the debates about, for instance, the Soviet Union. If one builds from a "free labor" vision of socialist transition, then work processes like those in the Soviet Union are *perfectly consistent* with that transitional view.

37. V. I. Lenin, "State and Revolution," in *Selected Works in One Volume* (New York, 1971), p. 337.

38. Aronowitz provides some of the citation for both Marx and Engels in "Marx, Technology, and Labor."

39. There is some related discussion of this issue in Frank Roosevelt, "Cambridge Economics as Commodity Fetishism."

40. The argument, in skeleton, builds upon a historical appreciation of the increasingly social character of peoples' lives— particularly under capitalism. It suggests that "free labor" is

inseparable from "collective labor," from the producers' point of view, because the latter provides the basis for the former. I do not intend the argument as a set of teleological propositions, but rather as a set of historical propositions. If there is a difference between the bourgeoisie's and the proletariat's (materially conditioned) viewpoints, surely this difference between "free labor" and "collective labor" constitutes a critical dimension of opposition.

41. There is a third aspect of capitalist development which some might add to this discussion. The tendency of the accumulation process to generate "crises of disproportionality" might prompt a rebellion against the "anarchy" of capitalist production and distribution. This contradiction has always seemed, at least to me, to provide the basis for little more than economic planning and certainly seems to provide an insufficient objective basis for a socialist revolution.

42. I use "fusion" here in the sense of L. Althusser in his essay on "Overdetermination" in *For Marx* (New York, 1969).

43. Our work on labor market segmentation is intended to provide the basis for much of this argument. See Gordon, Edwards, and Reich, "Labor Market Segmentation in American Capitalism," and Edwards, Reich and Gordon, eds., *Labor Market Segmentation* (Lexington, Mass., 1975).

Division of Labor in the Computer Field
by Joan Greenbaum

Throughout the 1960s there was intense academic debate about the effects of automation, particularly as represented by the computer, on the labor process. Bourgeois economists and sociologists, while admitting that automation frequently reduced skills among many kinds of workers, pointed to the growing employment in the computer industry itself as a bright spot for labor. Many of the new jobs in this growing field were categorized as technical and professional and were considered illustrative of labor-force upgrading.

Indeed, computer jobs were glorified in the early period. Salaries were high, and qualified computer technicians had a great deal of freedom and mobility. I entered the data-processing field almost 15 years ago in the heyday of its craft, and like many of my fellow workers rode the crest of its early opportunities. During those years, and on into the period of division of labor in the field, many of us tried to fight the changing conditions of our labor, but we lacked a conceptual base from which to present our arguments.

What was missing from these early evaluations was a firm understanding of the labor processes of capitalism. Marx's analysis is no less applicable to an occupation that could not have been conceived of in his day. In a short twenty-year span, work in the computer field has been transformed by capitalism to suit its needs, through carefully planned divsion of labor.

While economists, sociologists, and propagandists for computer usage have been so enamored of automation, it is no wonder that workers within the field have been blinded by their view of a growing high-skilled occupation. Harry Braverman, however, in his book *Labor and Monopoly Capital,* has at last separated the forest from the trees.

For a short time in the 1940s and early 1950s, the data-processing occupations displayed the characteristics of a craft. . . . The development of a data-processing craft was abortive, however, since along with the computer a new division of labor was introduced and the destruction of the craft greatly hastened. Each aspect of computer operations was graded to a different level of pay frozen into a hierarchy: systems managers, systems analysts, programmers, computer console operators, key punch operators, tape librarians, stock room attendants, etc. It soon became characteristic that entry into the higher jobs was at the higher level of the hierarchy, rather than through an all-around training. And the concentration of knowledge and control in a very small portion of the hierarchy became the key here, as with automatic machines in the factory, to control over the process.[1]

Starting from Braverman's overview, the purpose of the analysis presented here is not to bemoan the long-gone days of craft-like activity, but rather to highlight the course of events so that those of us in the computer field, and workers affected by it, can better grasp the implications.* The focus will be on those workers employed in computer-related jobs, specifically those jobs having to do with the processing of data. The computer field under study is made up of both computer-manufacturing companies, and the service bureaus, banks, and insurance firms heavily dependent on computer use. Although the largest user of computers has been the U.S. government and within it the military, this analysis will concentrate on commercial uses of computers in the office sector. In demonstrating the impact of division of labor on the labor process in the computer field, I hope that the illustrations will clarify the process of discipline used to reduce a largely "technical" work force to a highly segmentized "white collar" assembly line, where control of knowledge is concentrated. The tasks of programmers and computer operators will be explored here to offer examples of the transformation. The major themes include:

---

* I shall bend Braverman's time frame slightly as my evidence suggests that the real impetus for division of labor did not take hold until the mid-sixties. Subdivisions of computer workers existed during the earlier period, but the full effects were not felt by the workers until the industry expanded in the sixties.

(1) The division of labor and degradation of work already witnessed in the manufacturing sector, and to a growing degree in the service sector, has been compressed into a twenty-year time frame within the computer field. Its pattern differs only slightly from that drawn in other fields.

(2) The rapid growth of computer use initially created the urgent need for skilled workers. These workers were drawn away from other fields with the lure of high pay and job mobility. By the time the growth of computer use had begun to mature, there was a need to discipline workers to new productivity standards. Attempts to discipline workers played a major role in the movement toward standardization, routinization, and the downgrading of skilled tasks.

(3) While discipline played the major role in changing the work descriptions, technology was used to intensify productivity. In this, the most technologically based of areas, it appears that technology was not the *cause* of division of labor, but rather the battering ram to open the door to labor acquiescence. Indeed, changes in computer technology offered management the opportunity to speed information processing, cut personnel costs, and demand stricter standards from the workers.

(4) The rigid hierarchy was created to reinforce the effects of standardization. Job categories were minutely defined so that tasks could be performed at the lowest possible rate of pay. The resulting hierarchy reflected the class and race positions obtaining in society as a whole.

(5) The current crisis has driven smaller computer manufacturing companies and users from the field, resulting in increased centralization. The workers laid off from these firms, as well as the large supply of trained technicians turned out by schools, have formed a surplus labor force in the field. The effects on wages and further task reductions are now beginning to be felt.

(6) Trends in the immediate future seem to be toward lower salaries relative to the cost of living, expansion of clerical-like jobs, and a shift away from computer specialists. Technological skill has been removed from all but a handful of workers.

Early Computer Use: 1950-1965

The generally acknowledged start of the computer age dates from the early 1950s. As in the case of most technological birthdays there is no precise date; some say the installation of the first UNIVAC computer for use in the Bureau of the Census in 1951 marks the start of widespread commercial use, while others move the time to the entry of a UNIVAC computer for payroll processing at General Electric in 1954. It is clear that during the early 1950s Sperry Rand, the manufacturer of UNIVAC, had a "jump on the market," while IBM turned down a deal to acquire the rights to this machine "because it felt that the greatest market potential for computers was in scientific rather than business applications."[2] The clamor for commercial computers, coupled with the development of mass-production techniques in computer manufacturing during 1954-1955, seems to have convinced IBM of the error of its ways. "By making the most of Sperry Rand's mistakes,"[3] IBM prevented itself from becoming a brief footnote in history. In 1955 it overtook Rand for the computer manufacturing lead.

The first business computers were used for purely repetitive clerical functions which had previously been done by electric accounting machines. These systems, often seen as the predecessor to the modern computer, were based on electromechanical devices for processing keypunched cards. They evolved from keypunch card procedures developed by Herman Hollerith for the Bureau of the Census in 1890. While it is true that accounting functions (like payroll and accounts receivable) performed by the card-processing machines were absorbed by the faster electronic computers in the 1950s, the latter differed greatly from their mechanical ancestors. The most striking difference lay in the fact that the computers could be programmed by a series of changeable instructions, while the processes of the accounting machines were dependent on fixed actions initiated by a wired board. The programmable memory of the new computers allowed logical processes, such as routine decision-making functions, to be automated. The technological distinction created the need for a new breed of workers to program or instruct the new machines. Additionally, the increased equipment speed established the precedent for continuous processing, whereas the electric accounting machines (EAM) relied on multiple job steps involving worker intervention at each stage of processing.

By 1955 the industry was in urgent need of skilled personnel to operate, repair, and program the burgeoning computer applications. The workers who had previously operated the accounting machines could only partially fill this demand. It was not until the mid- to late-1960s, when computers were well entrenched, that EAM equipment began to be phased out. Until that time most companies maintained their accounting equipment in parallel operation with their newer computers. The tremendous demand by management for additional business information, and therefore control over information, led to a great increase in the demand for information processing, resulting in the use of both EAM equipment and computers simultaneously.

Lacking a large enough pool of pre-trained labor power, the computer manufacturers and users began to woo people away from the sciences, often offering them unlimited flexibility in their work, as well as comparatively high wages. Programmers, in particular, were like virtuosos in high demand who could jump from job to job, writing their own tickets to match their expectations. Almost all were quite young and sought independence and creativity in a field which promised status as well as a high pay.[4]

This system of pirating skilled labor power from other fields paid off for the industry until the widespread use of computers began to cause other pressures on capital. By 1962 there were 10,000 to 12,000 computers installed, employing about 150,000 workers in their manufacture, programming, operation, and maintenance.[5] While the workers were enjoying the effects of a seller's market for their labor power, management journals and marketing literature were beginning to call for standardization of job descriptions and routinization of tasks. High on the management list of reforms was an effort to stop the costly effects of personnel turnover, created by workers jumping to higher-paying data processing jobs. Certainly, the

50 percent growth in computer workers' salaries during the 1958-1962 period intensified the corporate drive to cut costs.⁶

Another argument put forth by management in favor of job restructuring stressed the fact that while early computer centers were often charged to Research and Development, by 1963 they were expected to pay their own way and show a "return on investment."⁷ The need for middle management to control the undisciplined, job-hopping work force, combined with the pressures from upper management to account for their expenditures, greatly hastened the death of craft-like worker activity.

Marx foresaw the imprint of capitalism on any industry when he wrote: "The separation of the intellectual powers of production from the manual labor, and the conversion of those powers into the might of capital over labor, is . . . finally completed by modern industry erected on the foundation of machinery."⁸

The computer field was no exception to this rule. During the 1950s three classes of technical workers were carved out: operators, the manual power to tend the machines, who were rapidly on their way to becoming "feeders" and "attendants"; programmers, the intellectual power to write the instructions and make the computer perform its tricks; and technicians, to repair the "engine" and watch over its functioning.* Although the trilogy existed by job classification in the 1950s, the workers themselves often overlapped tasks. It was not uncommon for programmers to forsake some of their more "intellectual" tasks in order to run the computer for the sheer pleasure of doing so. Similarly, operators, coming in contact with programmers in the computer room, would seize on the chance to learn what the programmers were doing, and thus enhance their understanding of programming and the possibilities for promotion. The computer center, or machine room, as it was appropriately

---

* A fourth category, that of data entry clerks, had previously been extracted from the technical labor pool, as its work patterns resembled those of other clerical workers, and therefore could not hide behind the mask of technical skill. See Braverman, *Labor and Monopoly Capital*, pp. 329-337, for a detailed treatment of routinization of clerical tasks among these workers.

called by the workers, was like a social hall, where the different categories of workers could meet and exchange techniques and ideas.

This initial division of labor was clearly not sufficient to meet the needs of management and capital. While workers were taunting management with their technical "expertise," management stepped up its drive for efficiency and what it saw to be its necessary co-requisite, division of labor. The independent computer labor force with its concentration and interchange of skills among the workers was clearly a threat to management. One loud cry from management was heard from Dick Brandon, an influential industry consultant, who argued that the industry had reached "economic maturity" without developing proper working methods, procedures, and disciplines. He called for tighter management controls, formal standards, and performance measurements, while decrying the "loss of management control" over data-processing functions.[9] Scientific management, or the systems approach as it was known in the computer field, took over, and the process of removing skills from each job classification was carefully executed.

Capital Takes Control: 1965-1970

Management demands for a more controllable work force were not the only impetus for change. Commercial uses of computers were catching on rapidly; and by 1965 IBM read the forecasts accurately and began to market the 360 computer system. The IBM 360 was the first general purpose computer designed to process the large volumes of data needed for business information processing. Although the technological base of this machine was not a marked departure from earlier models, it was promoted on its ability to speed the processing of information. The reasons for the success of the 360 could fill many volumes, for they range from IBM's self-held belief that the machine is technically superior, to its competitors' contention that IBM's 70-80 percent share of the computer manufacturing market insured success from a monopolistic point of view.[10] What is significant from the workers' viewpoint is that the larger machines promised greater efficiency and thus supported management's demands for greater worker productivity.

Those of us in the field at the time of the introduction of the 360 remember it well, for almost overnight a division of labor occurred, not by chance as it seemed to us then, but by clear design. One of the first management rulings to be enforced was a prohibition on programmers entering the computer room, thus isolating the two categories of labor, diminishing their social interaction, and cutting off the opportunity for the exchange of ideas. From a financial point of view, upper management saw the 360 as an increase in capital expenditure, requiring tighter controls and greater security for their new investment. Thus, to them, the separation of operators and programmers was a necessary step to control the computer room and protect their investment. Line management, on the other hand, was quick to respond to the need to divide the work force as an aid in their struggle to transfer some of the technical skill from the workers into their own domain.

Perhaps the best way to understand the changes that took place is to examine the functions of workers in the field. Programmers represented the pinnacle of intellectual job involvement and diversified skills. Basically, the programmer "directs the computer to do a job and provides it with detailed instructions—a program—as to how to accomplish the task. Thus, the programmer is, in a sense, an interpreter who, given a problem in science, engineering, or business, translates it into a form and language with which the computer can deal."[11]

A more glorified definition, and one often argued for by programmers themselves, states that they are "engaged in work which brings them to the brink of human knowledge," and that "the programmer must fulfill many positions in the course of solving a problem which include administrative know-how and scientific expertise."[12] A 1963 survey of programmers found that 88 percent had bachelors degrees while 30 percent had completed a masters.[13] The same study found that 75 percent of the programmers had a *highly* positive attitude toward their jobs, and emphasized that it was the diversity, challenge, and freedom of the work that accounted for this high degree of satisfaction.[14]

An early casualty in the movement to discipline this highly educated middle-class work force was the separation of analytical tasks from those that required only translation into programming. The job of systems analyst became the highest level in prestige and salary, as systems analysts were separated from programmers. Although both job descriptions still required technical expertise and thought processes, the systems analyst was to develop procedures to process information and determine the method and solution to business problems, while the programmer was to translate these solutions into a language the machine could understand.

The process of sorting out the repetitious tasks from those that might still require thought has been evolving during the last ten years. Programmers have been divided and subdivided into single-skill categories, and simplified programming languages have been developed as a means to concentrate skill into as few hands as possible.

A noteworthy movement in the codification of programmer tasks was the development over the last fifteen years of four levels of programming languages. Initially, programs had to be written in a detailed and complex format called machine language. This required a great deal of skill and knowledge of the machine, and was replaced by assembly language coding which simplified the instruction process by allowing the programmer to code fewer and less complicated instructions. By 1965 general purpose computers like IBM's 360 could use more generalized instruction sequences; and easier-to-use languages like COBOL, a language for business processing, came into widespread use. These languages removed the programmer from the technical detail of the equipment and required only the ability to transcribe a given solution into an English-like series of instructions. The last development has been the introduction of pre-planned application languages, where programmers need only insert a prearranged series of codes. The development of pre-planned applications has resulted in the total removal of technical skill from some of the tasks of programming. Starting in the late 1960s, more and more applications which would have previously been done by a data-processing worker have shifted out of the field. Accountants and bookkeepers, for example, code financial information directly into data terminals for transmission to the central computer center. These workers require minimal computer training compared to the programmer specialist of the early sixties.

The outcome is a hierarchical structure which requires a large number of coders whose task it is to take pre-planned specifications and code them into a proper format, and a few specialists who plan the application. Even the systems analysts who design the applications now produce their specifications within a framework of narrowly defined job steps.

During this process, as skill or craft work was abstracted from each task, programmers hid behind the myth of "professionalism" which kept them from organizing into unions and actually seeing the changes for what they were. Indeed, strong evidence suggests that the impetus to professionalism has come from management and not from the workers themselves. Trade journals and computer associations, both overwhelmingly management organs, have strongly pushed the concept.[15] It seems likely that programmers, feeling the tide of job degradation, only too gladly clung to the belief that they were professional. Computer operators would often boast that while programmers had status, they, the operators, made more money with their overtime pay included.

Operators were in some ways harder than programmers to mold. This was probably due to the fact that decisions and control over their work had already been removed from their jobs. Additionally, an "insubordinate" operator could easily wreak havoc on the daily operations of a computer room by slight infringements of the rules. An operator who mounted the wrong data tape for the processing of an accounting system, for example, could cause direct injury to the corporation's pocket. The effects of programming mistakes, on the other hand, were slower to come to notice and usually less costly.

There is also a noticeable class difference between the two job categories. While programmers were drawn from the college-educated middle class and were paid annual salaries, operations positions required only a high school diploma and were paid hourly wages comparable to factory workers. Typically, operators were young men of working-class families who saw the computer field as a step toward middle-class status, although this was never reflected in their wage structure or their job duties. As Braverman explains: "The training and education required for this job may perhaps best be estimated from the pay scales, which in the case of a Class A operator are on about the level of the craftsman in the factory, and for Class C operators on about the level of the factory operative."[16]

In terms of job duties, operators are involved in seeing that the throughput of data is properly loaded into the various pieces of computer machinery. Computer operators are, in fact, machine operators who load the equipment with data and press the appropriate controls for its processing. Early operators, like their programming counterparts, had to know the working of all parts of each machine and be familiar with a range of machines. But this type of job diversity and skill requirement was eliminated on the larger machines. The typical IBM 360 computer center saw the subdivision of operators into categories where each operator tended only one equipment component of a computer system. Thus data for the punched-card reader was handled by the "specialist" in charge of that device, and tape and disk devices were tended by what the trade called "tape jockeys." Console or main-panel operators required more skill, for they were to keep an eye on the total machine processing. Under almost no circumstances, however, were even they allowed to repair a machine part (the job of the technician) or to modify a program (the job of the programmer). After the mid-sixties, pay scales reflected the division of labor in the machine room so that each function was paid according to the lowest rate. Computer operators Class C, the beginners who manned the data devices, earned an average of $105 per week, while Class A operators, or those assigned to the console, could average about $150.[17] Although different wages for different experience levels had always been present, by the late sixties experience levels became firmly equated with specific machines, thus routinizing the tasks.

When it became apparent that it was difficult to bend the worker to further demands of the machines, the machines were recast to alleviate work steps. Again collaboration by upper and line management played a role. Since IBM advertised its tape drives as processing 156,000 characters per second, management became disenchanted with workers who would take several minutes to load a tape on a drive. Imagine the frustration of managers whose equipment was capable of processing the entire contents of a book in several seconds, when confronted with operators who failed to act with comparable speeds! In an effort to make the total machine processing more continuous, tape drives were introduced which would load magnetic tapes on devices automatically. Technically, the new devices did not represent a departure from the old. In fact the operation of tape drives had been similar to threading a home tape recorder, and the new modification made it resemble cassettes rather than individually wound reels.

It was during this period when skills were removed from both operations and programming jobs that the schools were beginning to turn out skilled computer workers. Universities were only too glad to respond to management pleas for more standardized programmers, and private institutes were quick to jump at the opportunity to make a profit with their matchbook and TV advertisements for computer operations training. Until the mid-sixties, management had been almost totally dependent on providing their own training, for few college or commercial training programs existed. It was the joint demands for performance standards and an increased number of computer workers that pushed the colleges to start computer science programs. Once begun, they churned out a large body of disciplined future workers, giving management a larger pool of labor power from which to choose.

Oversupply of Computer Workers: 1970-1975

The results of the processes outlined above were beginning to appear by the time the recession hit the computer field in 1971. The number of computer science graduates and technical training school students was beginning to flood the labor supply with potential talent beyond the level of skills required. A 1972 article in *Business Automation* recalls: "Remember the people problem? It was a prime topic of conversation through the 60's, but seemed to fade away in the past year or so. The job-hopping programmer or systems analyst will continue to be a rarity."[18] The article then goes on to outline the results, stating that (1) personnel costs are stabilizing, (2) turnover is slackening, and (3) supply is approaching demand. The following year, the director of the Association of Computer Programmers and Analysts noted that in addition to the growing number of available programmers, the skills required for applications programming were becoming less each year. He complained that "these pressures work together to lead some people to believe that programmers are being treated in a manner somewhat above their station. Instead of being treated like scientists, they should be treated more like bookkeepers."[19]

Indeed, programmers' salaries began to reflect this trend for entry level positions. Breaking a long-standing upward curve, the average starting salary for programmers stayed at $8,500 between 1970 and 1972.[20] To add insult to injury, a private survey conducted by the Robert Half Personnel Agency, found that overall programmers' salaries in large installations rose only 2 percent in 1975 over the previous year.[21] It should be added that large computer installations usually showed the higher salaries and most growth.

Despite the flood of articles discussing this turnaround in programmer supply, schools and career counselors have continued the production of the now surplus programming population. Perhaps the best summary of the changing events was offered by A.P. Ershov, head of the Information Division of the USSR Academy of Sciences. Writing with an outsider's perspective about the U.S. situation he said:

> The volume of work to be done is increasing, and wages less so. The romantic aura surrounding this inscrutable occupation is, if it ever really existed, beginning to fade. . . . Even the claim of programmers to be a special breed of professional employee has come to be disputed. Still more significant, authority over the freewheeling brotherhood of programmers is slipping into the paws of administrators and managers—who try to make the work of programmers planned, measurable, uniform, and faceless.[22]

In addition to the impact of stricter management control, and increased number of trained entry-level workers, the 1970s have witnessed a movement toward greater centralization among firms engaged in data processing. Starting with the recession in 1971, many of the smaller service bureaus have closed their doors, leaving computer workers on the job market. Even huge corporations like RCA and Honeywell have given up the struggle against IBM and closed their computer manufacturing divisions. As the less-capitalized firms give up to their bigger competitors, the larger data processing establishments are centralizing their operations for greater efficiency. Whereas a corporation may have previously maintained several computer centers for different business functions, today the push is toward one central facility with remote "time-sharing" terminals in corporate offices around the country. The advent of remote terminals means, in effect, that any desk within reach of a telephone can transmit data over phone wires to a distant computer. Retail stores have begun to use these services for up-to-the-minute inventory reports, as clerks key in information on each sale. Banks, insurance companies, credit-reporting firms, and supermarkets are all tapping the new potential. The introduction of automated supermarket checkout counters this year is expected to break new ground in large-scale use.

The results for computer workers are, of course, major. A central computer facility for each company, or in many cases, groups of companies, requires fewer operators and programmers. And while some of the more sophisticated remote terminals need computer operations personnel, these devices are relatively simple to run. A recent article in *Computerworld,* the computer industry's weekly newspaper, prophesied that these trends indicate the industry will need only a few "hot shot" programmers, for users are now performing their own data processing via terminals.[23] Computer operators appear to face the same fate. The 1975 salary survey mentioned earlier reported no increase in operators' salaries between 1974 and 1975, indicating that this was due to the "greater influx of entry-level people."[24]

It is beyond the scope of this analysis to project the impact of these trends on workers outside the data processing field. Data processing operators and programmers will still be needed to meet the needs of expanded computer use. Their tasks, however, have changed over the last decade, and it appears that their salaries are also reflecting these changes. The *Computer Manpower Outlook*, published in 1974 by the Department of Labor states: "Employment in computer occupations is expected to grow more slowly over the 1970-1980 period than during the past decade, and the distribution of workers among computer jobs is expected to change."[25] Citing labor costs and technological change as the prime ingredients in this process, the authors argue:

> Because costs of computer manpower are a major part of computer user costs, manufacturers have a strong incentive to reduce the manpower needed to use their equipment by incorporating functions that currently are being performed by computer personnel into the hardware [equipment]. Also technological innovations that enable workers in other occupations to interact directly with computers and thus eliminate costly data processing specialists are expected to be stressed.[26]

It is interesting to note that steps toward job degradation, begun little more than a decade ago, have so rapidly produced results. The division of labor among computer workers has relieved them unwillingly of their skills and left increasing numbers of them high and dry on the labor market.

Notes

1. Harry Braverman, *Labor and Monopoly Capital* (New York, 1974), p. 329.
2. "Survey and Study of the Computer Field," *Computers and Automation*, January 1963, p. 23.
3. Ibid.
4. "A Profile of the Programmer," a study by Deutsch and Shea, Inc., New York, *Industrial Relations News*, 1963, pp. 8-31.
5. "Survey and Study of the Computer Field," p. 23.
6. Dick Brandon and Fredrich Kirch, "The Case for D.P. Standards," *Computers and Automation*, November 1963, pp. 28-31.
7. Ibid.

8. Karl Marx, *Capital*, Vol. I, (New York, 1967, 1974), p. 423.
9. "The Case for D.P. Standards," p. 29.
10. See Gerald W. Brock, *The U.S. Computer Industry: A Study in Market Power* (Cambridge, Mass., 1975), for an extensive discussion of IBM's market power.
11. "A Profile of the Programmer," p. 3.
12. A.P. Ershov, "Aesthetics and the Human Factor in Programming," *Communications of the ACM*, Vol. 15, no. 7 (July 1972), pp. 503-504.
13. "A Profile of the Programmer," p. 21.
14. Ibid., p. 6.
15. Philip Kraft, "Pushing Professionalism, Programming the Programmer," *Science for the People*, Vol. 3, no. 4 (July 1974), pp. 26-29.
16. Braverman, *Labor and Monopoly Capital*, p. 330.
17. "1967 Report, EDP Jobs and Salaries," *Business Automation*, June 1967, pp. 40-49.
18. "The 70's: Retrospect and a Look Ahead," *Business Automation*, January 1972.
19. T.D.C. Kuch, "Unions or Licensing? or Both or Neither," *Infosystems*, January 1973, p. 42.
20. *Occupational Outlook Handbook*, 1970-71 edition, 1973-74 edition, U.S. Department of Labor, Bureau of Labor Statistics.
21. *Computerworld*, September 3, 1975, p. 4.
22. "Aesthetics and the Human Factor in Programming," p. 501.
23. "Remote Maintenance: Trend of the Future," *Computerworld*, August 21, 1974, pp. 1-2.
24. *Computerworld*, September 3, 1975, p. 4.
25. *Computer Manpower Outlook*, U.S. Department of Labor, Bureau of Labor Statistics, Bulletin 1826 (1974), p. 3.
26. Ibid.

Marx as a Student of Technology
by Nathan Rosenberg

I

This paper will attempt to demonstrate that a major reason for the fruitfulness of Marx's framework for the analysis of social change was that Marx was, himself, a careful student of technology. By this I mean not only that he was fully aware of, and insisted upon, the historical importance and the social consequences of technology. That much is obvious. Marx additionally devoted much time and effort to explicating the distinctive characteristics of technologies, and to attempting to unravel and examine the inner logic of individual technologies. He insisted that technologies constitute an interesting subject, not only to technologists, but to students of society and social pathology as well, and he was very explicit in the introduction of technological variables into his arguments.

I will argue that, quite independently of whether Marx was right or wrong in his characterization of the future course of technological change and its social and economic ramifications, his formulation of the problem still deserves to be a starting point for any serious investigation of technology and its ramifications. Indeed, the following statement by Marx, amazingly fresh over a century later, reads like a prolegomenon to a history of technology which still remains to be written:

> A critical history of technology would show how little any of the inventions of the eighteenth century are the work of a single individual. Hitherto there is no such book. Darwin has interested us in the history of Nature's Technology, i.e.,

---

In preparing this paper, I have had the considerable benefit of comments and suggestions from Jens Christiansen, Paul David, David Mowery, and Stanley Engerman. I am grateful also to the National Science Foundation for financial support.

in the formation of the organs of plants and animals, which organs serve as instruments of production for sustaining life. Does not the history of the productive organs of man, of organs that are the material basis of all social organization, deserve equal attention? And would not such a history be easier to compile since, as Vico says, human history differs from natural history in this, that we have made the former, but not the latter? Technology discloses man's mode of dealing with Nature, the process of production by which he sustains his life, and thereby also lays bare the mode of formation of his social relations, and of the mental conceptions that flow from them.[1]

In what follows, I will focus first upon Marx's alleged technological determinism, second on his views on the characteristics of Modern Industry which are responsible for its high degree of technological dynamism, and finally on the special importance which Marx attaches to the role of the capital-goods sector in the generation of technological change. An important question regarding Marx's treatment of technology is, quite simply, what it was about his method or approach which made him so much more perceptive on this subject than any of his contemporaries. I think a few tentative methodological observations, upon which I will elaborate later, may be useful at the outset.

The method of historical materialism which Marx utilized was one which emphasized the interactions and conflicts of social classes and institutions, not individuals. Thus for Marx invention and innovation, no less than other socioeconomic activities, were best analyzed as social processes rather than as inspired flashes of individual genius. The focus of Marx's discussion of technological change is thus not upon individuals, however heroic, but upon a collective, social process in which the institutional and economic environments play major roles.

Marx's historical approach was one which emphasized the discontinuous nature of social evolution, an evolutionary process which was, for him, moving forward under capitalism, just as it had under earlier forms of social organization. Rather than viewing capitalism as the final, logical outcome of a smooth, lengthy evolutionary process, Marx treated it as simply one stage in the process of historical evolution, while looking for the unique features of the capitalist forces of production and attempting to understand them in dynamic terms.

I would argue that the dialectical method is the most important factor in understanding the methodological basis for Marx's unique insights. Rather than positing some unidirectional chain of causation for technological change, Marx offers a far richer mode of analysis, one which emphasizes the mutual interactions and feedbacks between economy and technology. His analysis of the rise of the system of "machinofacture" and its implications for technological change, which I will examine in this paper, is an important example of the insights yielded by such a method.

II

First of all, it is necessary to advert briefly to the oft-repeated view that Marx was a technological determinist.[2] If by this we mean that technological forces are the decisive factor in generating socioeconomic changes—that technological factors are, so to speak, the independent variable in generating social change, which constitutes the dependent variable—it is easy to demonstrate that Marx subscribed to no such simplistic view. No doubt certain passages can be cited to support such an interpretation—most notably, of course, the statement from *The Poverty of Philosophy* that "The handmill gives you society with the feudal lord; the steam-mill, society with the industrial capitalist."[3] But such an interpretation, relying upon a few such aphoristic assertions, often tossed out in the heat of debate (as the quotation from *The Poverty of Philosophy* was thrown out in criticizing Proudhon) finds little support in Marx's own treatment of the major historical episodes with which he was concerned. Indeed, in a process no less central to Marx than the historical rise of capitalism itself, technological factors play no immediate role at all. For Marx, capitalism developed in Western Europe basically in response to growing markets and related opportunities for profit-making associated with the geographic explorations of the fifteenth century. The *locus classicus* for this view is, of course, the opening pages of the *Communist Manifesto*:

> The discovery of America, the rounding of the Cape, opened up fresh ground for the rising bourgeoisie. The East-Indian and Chinese markets, the colonization of America, trade with the colonies, the increase in the means of exchange and in commodities generally, gave to commerce, to navigation, to industry, an impulse never before known, and thereby, to the revolutionary element in the tottering feudal society, a rapid development.
>
> The feudal system of industry, under which industrial production was monopolized by closed guilds, now no longer sufficed for the growing wants of the new markets. The manufacturing system took its place. The guild-masters were pushed on one side by the manufacturing middle class; division of labor between the different corporate guilds vanished in the face of division of labor in each single workshop.
>
> Meantime the markets kept ever growing, the demand ever rising. Even manufacture no longer sufficed. Thereupon, steam and machinery revolutionized industrial production. The place of manufacture was taken by the giant, Modern Industry, the place of the industrial middle class, by industrial millionaires, the leaders of whole industrial armies, the modern bourgeois.
>
> Modern industry has established the world-market, for which the discovery of America paved the way. This market has given an immense development to commerce, to navigation, to communication by land. This development has, in its turn, reacted on the extension of industry; and in proportion as industry, commerce, navigation, railways extended, in the same proportion the bourgeoisie developed, increased its capital, and pushed into the background every class handed down from the Middle Ages.[4]

I have taken the liberty of quoting at length from a familiar source because this passage is, it seems to me, a definitive refutation of the view that Marx was a technological determinist. The question is: What are the factors which *initiate* change? What are the factors which cause other factors to change? A resolute technological determinist would presumably argue that the entire process of European expansion was initiated by navigational improvements which in turn generated the growth of overseas markets. But this is clearly not Marx's own view here. In fact, Marx states *twice* that the improvements in navigation were caused by the prior growth in markets and commercial opportunities. The passage makes it unmistakably clear that the technological changes associated with the two stages of capitalist development—the manufacturing system and Modern Industry—were responses to an expanding universe of profit-making opportunities.

For Marx, then, capitalist relationships emerged when the growth of profit-making opportunities led to an expansion in the size of the productive unit beyond that which was characteristic of the medieval craft workshop. The mere quantitative expansion of such workshops led eventually to qualitative changes of a most basic sort in social relationships.[5] Although the system of manufacture totally dominated the first two and a half centuries of Western capitalism and led to major transformations in social relationships,[6] it was not associated with any major technological innovations. "With regard to the mode of production itself, manufacture, in its strict meaning, is hardly to be distinguished, in its earliest stages, from the handicraft trades of the guilds, otherwise than by the greater number of workmen simultaneously employed by one and the same individual capital. The workshop of the mediaeval master handicraftsman is simply enlarged."[7]

To regard Marx as a technological determinist, then, is tantamount to ignoring his dialectical analysis of the nature of historical change.[8] The essence of this view is that the class struggle, the basic moving force of history, is itself the product of fundamental contradictions between the forces of production and the relations of production. At any point in historical time, new productive forces emerge, not exogenously or as some mysterious *deus ex machina*, but rather as a dialectical outcome of a larger historical process in which *both* the earlier forces and relations of production play essential roles. As Marx forcefully put it: "It must be kept in mind that the new forces of production and relations of production do not develop out of *nothing*, nor drop from the sky, nor from the womb of the self-positing Idea; but from within and in antithesis to the existing development of production and the inherited, traditional relations of property."[9]

Thus, for Marx the basic rhythm of human history is the outcome of this dialectical interaction between the forces and the relations of production. To categorize Marx as a technological determinist one would have to demonstrate first that he does *not* intend his historical argument to proceed in a dialectical form. I think it is easy to demonstrate that he does.

III

If Marx was not a technological determinist he did, nonetheless, attach great importance to technological factors. The reasons for this are made clear in Chapter 7 of the first volume of *Capital*. Technology is what mediates between man and his relationship with the external, material world. But in acting upon that material world, man not only transforms it for his own useful purposes (that is to say, "Nature becomes one of the organs of his activity"[10]) but he also, unavoidably, engages in an act of self-transformation and self-realization. "By thus acting on the external world and changing it, he at the same time changes his own nature."[11] Technology, therefore, is at the center of those activities which are distinctively human. For technology comprises those instruments which determine the effectiveness of man's pursuit of goals which are shaped not only by his basic instinctive needs, but also those formulated and shaped in his own brain. "A spider conducts operations that resemble those of a weaver, and a bee puts to shame many an architect in the construction of her cells. But what distinguishes the worst architect from the best of bees is this, that the architect raises his structure in imagination before he erects it in reality. At the end of every labor process, we get a result that already existed in the imagination of the laborer at the commencement."[12]

The informed student of society, therefore, can infer much concerning the nature of a society, its intellectual attainments, its organization, and its dominant social relationships, by studying the instruments of human labor. But again it needs to be insisted that Marx's position here cannot be reduced to a crude technological determinism. In a highly perceptive passage which is sometimes cited as evidence of his technological determinism, Marx is, in fact, pointing to what can be *inferred* about the nature of earlier societies from their remaining artifacts. "Relics of by-gone instruments of labor possess the same importance for the investigation of extinct economical forms of society, as do fossil bones for the determination of extinct species of animals. It is not the articles made, but how they are made, and by what instruments, that enables us to distinguish different economical epochs. Instruments of labor not only supply a *standard* of the degree of development to which human labor has attained, but they are also *indicators* of the social conditions under which that labor is carried on."[13] I believe this passage is evidence of technological determinism in about the same sense that one is entitled to say that a thermometer *determines* body temperature or a barometer *determines* atmospheric pressure. All such statements are equally specious and misleading in mixing up measurement with causation.

The decisive technological changes with which Marx is concerned begin around the middle of the eighteenth century. It is at this point that Britain began her transition from an industrial system of manufacture to what Marx calls Modern Industry. The general outline of his analysis, as presented in Chapters 13-15 of *Capital*, is well known and need not be repeated here. Rather, I will focus on certain specific features which, I believe, are still insufficiently appreciated.

First of all, Marx posed and dealt with a basic question concerning the nature of technology which has never received the attention it deserves. It is widely accepted that modern capitalist societies have achieved high levels of productivity because of the systematic application of scientific knowledge to the productive sphere. As Kuznets has stated: "The epochal innovation that distinguishes the modern economic epoch is the extended application of science to problems of economic production."[14] Our awareness of the importance of modern science (an awareness which is not, of course, confined to its purely economic significance) has led to a mushrooming interest in the history of science, which is now a thoroughly respectable academic discipline. But although we study the history of science (in some cases with the financial support of the National Science Foundation), the study of the history of technology is still largely (although by no means entirely) neglected. And yet to the extent that we are interested in the economic importance of science, we need to study the history of technology, because not all technologies will permit, or will permit in equal degrees, the *application* of scientific knowledge to the productive sphere. The growth of science, *by itself,* is not a sufficient condition for the growth of productivity. To believe that it is, is to ignore the mediating role of technology between man and nature. It was one of Marx's most important accomplishments to have posed precisely this question: What are the characteristics of technologies which make it possible to apply scientific knowledge to the productive sphere? Moreover, I think Marx suggested an answer which was quite adequate for his own historical period, given the nature of the industrial technology of his day and the state of scientific development. Still, this immensely important question needs to be posed and studied anew for the far more complex technologies, as well as the far more sophisticated bodies of scientific knowledge, which have emerged in the century since Marx wrote *Capital.* From the vantage point of the mid-nineteenth century, Marx's answer ran along the following lines.

The manufacturing system, which was the dominant mode of production of early capitalism, developed a high degree of worker specialization. Whereas the medieval handicraftsmen performed a whole range of operations in the production of a single commodity, the manufacturing system broke down the productive process into a series of discrete steps, and assigned each step to a separate detail laborer. However, although this growing specialization of work had highly significant consequences with which Marx was intensely concerned, it nevertheless shared a basic feature in common with the medieval handicraft system: a continued reliance upon human skills and capacities.

> Whether complex or simple, each operation has to be done by hand, retains the character of a handicraft, and is therefore dependent upon the strength, skill, quickness, and sureness, of the individual workman in handling his tools.

The handicraft continues to be the basis. This narrow technical basis excludes a really scientific analysis of any definite process of industrial production, since it is still a condition that each detail process gone through by the product must be capable of being done by hand and of forming, in its way, a separate handicraft. It is just because handicraft skill continues, in this way, to be the foundation of the process of production, that each workman becomes exclusively assigned to a partial function, and that for the rest of his life, his labor power is turned into the organ of this detail function.[15]

Although, therefore, the manufacturing system achieved a growth in productivity through the exploitation of a new and more extensive division of labor, a rigid ceiling to the growth in productivity continued to be imposed by limitations of human strength, speed, and accuracy. Marx's point, indeed, is more general: science itself can never be extensively applied to the productive process so long as that process continues to be dependent upon forces the behavior of which cannot be predicted and controlled with the strictest accuracy. Science, in other words, must incorporate its principles in impersonal machinery. Such machinery may be relied upon to behave in accordance with scientifically established physical relationships. Science, however, cannot be incorporated into technologies dominated by large-scale human interventions, for human action involves too much that is subjective and capricious. More generally, human beings have wills of their own and are therefore too refractory to constitute reliable, i.e., controllable, inputs in complex and interdependent productive processes.

The decisive step, then, was the development of a machine technology which was not heavily dependent upon human skills or volitions, where the productive process was broken down into a series of separately analyzable steps. The historic importance of the manufacturing system was that it had provided just such a breakdown. The historic importance of Modern Industry was that it incorporated these separate steps into machine processes to which scientific knowledge and principles could now be routinely applied. "The principle, carried out in the factory system, of analyzing the process of production into its constituent phases, and of solving the problems thus proposed by the application of mechanics, of chemistry, and of the whole range of the natural sciences, becomes the determining principle everywhere."[16] When this stage has been reached, Marx argues, technology becomes, for the first time, capable of indefinite improvement:

> Modern industry rent the veil that concealed from men their own social process of production, and that turned the various, spontaneously divided branches of production into so many riddles, not only to outsiders, but even to the initiated. The principle which it pursued, of resolving each process into its constituent movements, without any regard to their possible execution by the hand of man, created the new modern science of technology. The varied, apparently unconnected, and petrified forms of the industrial processes now resolved themselves into so many conscious and systematic applications of natural science to the attainment of given useful effects. Technology also discovered the few main fundamental forms of motion, which despite the diversity of the instruments used, are necessarily taken by every productive action of the human body; just as the science of mechanics sees in the most complicated machinery nothing but the continual repetition of the simple mechanical powers.[17]

I suggest that Marx's insight into the historical interrelationships between science and technology was extraordinarily perceptive and that it ought to be treated as a starting point for the vastly more complex interrelationships which have characterized the last century of capitalist development.[18]

IV

In the remainder of this paper I propose to concentrate upon Marx's analysis of the unique role and importance of the capital-goods sector. Although Marx is generally recognized as the father of the two-sector model, this recognition has been confined primarily to the usefulness of such models in the explanation of the inherent instability of capitalist economies, after the fashion in which Marx employed such models in the second volume of *Capital*. His work, however, suggests much more than this. The identification and isolation of a capital-goods producing sector offers rich possibilities for the further understanding of the form and the mechanism of diffusion of technological change, as well as other critical aspects of the behavior of capitalist societies.

In the early stages of the development of Modern Industry, machinery was produced by handicraft and manufacturing methods.[19] Although such methods were sufficient in the early stages, the growth in machine size and complexity and requirements for improvement in machine design and performance characteristics created demands which proved to be incompatible with the limited capacities of handicraft and manufacturing technologies. Indeed, "Such machines as the modern hydraulic press, modern powerloom, and the modern carding engine, could never have been furnished by Manufacture."[20] The realization of the full productive possibilities of Modern Industry therefore required that machine techniques be employed in the construction of the machines themselves. This is the final stage in the "bootstrap" operation, the stage by which Modern Industry completes its liberation from the constraints of the old technology. "Modern Industry had therefore itself to take in hand the machine, its characteristic instrument of production, and to construct machines by machines. It was not till it did this, that it built up for itself a fitting technical foundation, and stood on its own feet. Machinery, simultaneously with the increasing use of it, in the first decades of this century, appropriated, by degrees, the fabrication of machines proper."[21]

Once this stage of technological maturity has been attained, modern capitalism may be regarded as being in full possession of those extraordinary technological means which sharply distinguish it from all earlier stages in the development of man's productive capacities:

> So soon . . . as the factory system has gained a certain breadth of footing and a definite degree of maturity, and, especially, so soon as its technical basis, machinery, is itself produced by machinery; so soon as coal mining and iron mining, the metal industries, and the means of transport have been revolutionized; so soon, in short, as the general conditions requisite for production by the modern industrial

system have been established, this mode of production acquires an elasticity, a capacity for sudden extension by leaps and bounds that finds no hindrance except in the supply of raw material and the disposition of the produce.[22]

It was one of Marx's enduring accomplishments that he was among the first to perceive the inevitability of the trend toward bigness. This perception was, again, firmly rooted in his careful study of technological forces at work in mid-nineteenth century British capitalism. Marx asserted in Chapter 25 of the first volume of *Capital* ("The General Law of Capitalist Accumulation") the decisive economic advantages of capitalist production on a large scale. The nature of these advantages is carefully categorized and analyzed, and numerous specific examples are presented in Chapter 5 of the third volume ("Economy in the Employment of Constant Capital").

When, as a result of the process of capital accumulation, the capitalist economy has acquired a sufficiently large complement of capital goods, and therefore also a well-defined sector devoted to the production of capital goods, the system at this stage acquires a new source of productive dynamism. First of all there are the opportunities, when the scale of production is sufficiently large, for the exploitation of what we have come to call indivisibilities:

> In a large factory with one or two central motors the cost of these motors does not increase in the same ratio as their horse power and, hence, their possible sphere of activity. The cost of the transmission equipment does not grow in the same ratio as the total number of working machines which it sets in motion. The frame of a machine does not become dearer in the same ratio as the mounting number of tools which it employs as its organs, etc. Furthermore, the concentration of means of production yields a saving on buildings of various kinds not only for the actual workshops, but also for storage, etc. The same applies to expenditures for fuel, lighting, etc. Other conditions of production remain the same, whether used by many or by few.[23]

Furthermore, when production takes place on a sufficiently large scale, it eventually becomes worthwhile to take steps to utilize waste, or by-product, materials. "The general requirements for the reemployment of these 'excretions' are: large quantities of such waste, such as are available only in large-scale production, improved machinery whereby materials, formerly useless in their prevailing form, are put into a state fit for new production; scientific progress, particularly chemistry, which reveals the useful properties of such waste."[24]

Most generally, the existence of a large stock of capital goods now provides powerful economic incentives for innovations of a capital-saving nature. That is to say, there now exist great opportunities for increasing the rate of profit "by reducing the value of the constant capital required for commodity production."[25] This involves not only the development of improved machinery—steam engines which deliver a greater amount of power with the same expenditure of capital and fuel—but also *technological change in the machinery-producing sector itself.* As a result of such improvements in the machine-building sector, "although the value of the fixed portion of constant capital increases continually with the development of labor on a large scale, it does not increase at the same rate."[26] At this most advanced stage of capitalist development, inter-industry relationships come to play a most important role, because the rate of profit in one industry now becomes a function of labor productivity in another industry. "What the capitalist thus utilizes are the advantages of the entire system of the social division of labor. It is the development of the productive power of labor in its exterior department, in that department which supplies it with means of production, whereby the value of the constant capital employed by the capitalist is relatively lowered and consequently the rate of profit is raised."[27]

From an even broader perspective, the rate of profit may be increased by any capital-saving innovations, in whatever form. A large class of such innovations will therefore include all measures which reduce the turnover period of capital. From an economy-wide point of view, this has been precisely the effect of the communications revolution, one of the effects of which was a drastic reduction in the requirements for circulating capital. Marx deals with this source of capital-saving in Chapter 4 of the third volume of *Capital* ("The Effect of the Turnover on the Rate of Profit"):

The chief means of reducing the time of circulation is improved communications. The last 50 years have brought about a revolution in this field, comparable only with the industrial revolution of the latter half of the eighteenth century. On land the macadamized road has been displaced by the railway, on sea the slow and irregular sailing vessel has been pushed into the background by the rapid and dependable steamboat line, and the entire globe is being girdled by telegraph wires. The Suez Canal has fully opened East Asia and Australia to steamer traffic. The time of circulation of a shipment of commodities to East Asia, at least twelve months in 1847 . . . , has now been reduced to almost as many weeks. The two large centers of the crises of 1825-1857, America and India, have been brought from 70 to 90 percent nearer to the European industrial countries by this revolution in transport. . . . The period of turnover of the total world commerce has been reduced to the same extent, and the efficacy of the capital involved in it has been more than doubled or trebled. It goes without saying that this has not been without effect on the rate of profit.[28]

All this from someone who is often described as treating the process of technological innovation as if it were purely a labor-saving phenomenon!

One further point which deserves to be made here is that Marx does not regard the reliance upon a new power source as the crucial element in the development of machines. Indeed, as he points out, in many of the early machines man was himself the prime mover. The distinctive element, as our earlier discussion has suggested, is the transfer of the control over the tool out of human hands.[29] This transfer of control involves a quantum leap forward, since "the number of tools that a machine can bring into play simultaneously is from the very first emancipated from the organic limits that hedge in the tools of a handicraftsman."[30] Furthermore, modern technology has even, *mirabile dictu*, invented a substitute for the human hand itself in the form of Henry Maudsley's slide rest.[31] This simple but ingenious device, as Marx perceptively notes, replaces not any particular tool, "but the hand itself." In this sense the slide rest is a technological breakthrough fully comparable in importance to the steam engine.

In his discussion of technological change within the capital goods sector, Marx has many acute observations which are tossed out without further development. New inventions often contain inefficient design features at the outset because the inventor has not shaken himself totally free of an earlier technology which is being displaced and whose operating principles have been rendered irrelevant. Marx observes:

> To what an extent the old forms of the instruments of production influenced their new forms at first starting, is shown by, amongst other things, the most superficial comparison of the present powerloom with the old one, of the modern blowing apparatus of a blast-furnace with the first inefficient mechanical reproduction of the ordinary bellows, and perhaps more strikingly than in any other way, by the attempts before the invention of the present locomotive, to construct a locomotive that actually had two feet, which after the fashion of a horse, it raised alternately from the ground. It is only after considerable development of the science of mechanics, and accumulated practical experience, that the form of a machine becomes settled entirely in accordance with mechanical principles, and emancipated from the traditional form of the tool that gave rise to it.[32]

It is a shame that posterity has been deprived of some sketch of the ill-fated, two-footed locomotive! But, more seriously, the last sentence can be read as a striking anticipation of some of the central ideas of Abbott Payson Usher, probably the most careful twentieth-century student of the history of technology.[33] For Marx is insistent that technology has to be understood as a social process. The history of invention is, most emphatically, not the history of inventors. Here, as in so many other realms, Marx's position cannot be understood without dealing with the basic methodological question: What is the most appropriate unit of analysis? His answer is that for questions pertaining to long-term changes in technology, the individual is *not* the appropriate unit. In this particular instance Marx is insisting that technological change cannot be adequately understood by examining the contributions of single individuals, even though he often acknowledges the noteworthy contributions of such individuals. Rather, one needs to examine the way in which larger social forces continually alter the focus of technological problems which require solutions. Within this framework one may then examine how the productive process has, in the past, shaped the development of scientific and technological knowledge and skills.[34] One is then in a position to explore the social process of problem formulation and eventual solution. In all of this, however, although individual human beings are, inevitably, the actors, the *dramatis personae* of the historical process, the actual unfolding of the plot turns upon the larger social forces which shape their actions. Within this larger framework it is possible to see the efforts and contributions of *numerous* individuals even though those twin bastions of individualism, the patent office and writers of history textbooks, require that the names of *single* individuals be written alongside the names of particular inventions. But what is really involved is a process of a cumulative accretion of useful knowledge, to which many people make essential contributions, even though the prizes and recognition are usually accorded to the one actor who happens to have been on the stage at a critical moment.

Usher would agree with Marx's stricture, quoted earlier, that "a critical history of technology would show how little any of the inventions of the eighteenth century are the work of a single individual." Usher's own analysis turns upon the study of problem formulation in dealing with technology, focusing especially upon the process which he describes as "the setting of the stage." It is the correct setting of the stage which makes possible the eventual act of insight leading to the solving of the problem (whether Watt's separate condenser or Bessemer's converter). As Usher points out: "Our analysis . . . redefines the question. It is not necessary to explain the final act of insight; the task now consists in explaining how the stage is set to suggest the solution of the perceived problem."[35] Moreover, this is not the end of the process, because the initial act of insight is likely to lead to crude and primitive solutions. "The setting of the stage leads directly to the act of insight by which the essential solution of the problem is found. But this does not bring the process to an end. Newly perceived relations must be thoroughly mastered, and effectively worked into the entire context of which they are a part. The solution must, therefore, be studied critically, understood in its fullness, and learned as a technique of thought or action. This final stage can be described as critical revision."[36]

I would regard the completion of Usher's stage of critical revision as bringing us to essentially the same point in the development of a technology that Marx had in mind in referring to the process by which "the form of a machine becomes settled entirely in accordance with mechanical principles, and emancipated from the traditional form of the tool that gave rise to it." Marx was aware of the regular need for something resembling Usher's final stage of critical revision and attached considerable importance to it. Indeed, there is implicit in Marx's analysis a kind of "life cycle" in the development of new techniques of production. New machines, when first introduced, are usually economically inefficient for two quite distinct reasons. First of all, the initial model has not yet had the opportunity to be subjected to a rigorous examination of its operations from which methods for performance improvement can be expected to flow. Marx pointed to continual improvements

> which lower the use-value, and therefore the value, of existing machines, factory buildings, etc. This process has a particularly dire effect during the first period of newly introduced machinery, before it attains a certain stage of maturity, when it continually becomes antiquated before it has time to reproduce its own value. This is one of the reasons for the flagrant prolongation of the working time usual in such periods, for alternating day and night shifts, so that the value of the machine may be reproduced in a shorter time without having to place the figures for wear and tear too high. If, on the other hand, the short period in which the machinery is effective (its short life vis-à-vis the anticipated improvements) is not compensated in this manner, it gives up so much of its value to the product through moral depreciation that it cannot even compete with hand-labor.[37]

The early model, therefore, is recognized to have a short life expectancy—a high rate of "moral depreciation"—and this expectation that it will shortly be swept away by the competition of improved models is an ever present consideration in the mind of the capitalist.

Eventually the new machine, having been subjected to a series of design improvements, assumes a relatively stabilized form, and at this point it becomes possible for the capital-goods sector to develop techniques for producing the machine more cheaply. This is where the capital-goods sector plays its critical role in the ongoing competitive process:

> After machinery, equipment of buildings, and fixed capital in general, attain a certain maturity, so that they remain unaltered for some length of time at least in their basic construction, there arises a similar depreciation due to improvements in the methods of reproducing this fixed capital. The value of the machinery, etc., falls in this case not so much because the machinery is rapidly crowded out and depreciated to a certain degree by new and more productive machinery, etc., but because it can be reproduced more cheaply. This is one of the reasons why large enterprises frequently do not flourish until they pass into other hands, i.e., after their first proprietors have been bankrupted, and their successors, who buy them cheaply, therefore begin from the outset with a smaller outlay of capital.[38]

Marx emphasized in several other places the high cost of the early machine models by comparison with the later ones.[39] His argument suggests a great deal, not only about the process through which a capitalist economy generates new techniques, but also about the speed with which new techniques will be spread throughout the economy.[40]

I do not propose to discuss the question whether Usher's work was influenced by Marx. That is not, in my view, terribly important. But I do want to insist upon the fruitfulness of certain ways of conceptualizing the technological process under capitalism which Marx suggested but never developed beyond some precocious and suggestive hints. I want also to render my judgment that American students of Marx, in creating a mode of analysis which they call "the Marxist tradition," have not been faithful, in at least one important respect, to the mode of analysis initiated by Marx himself. For Marx, as I have argued, was a close student both of the history of technology and its newly emerging forms.

These strictures are only somewhat less applicable to the British Marxian tradition. For although it is true that the two leading British figures in the history of technology, J.D. Bernal and Joseph Needham, have frequently adverted to the strong Marxian component of their thinking, they have not, any more than others, followed Marx's hints or elaborated upon his insights concerning the development of modern industrial technology. This is a task which still remains to be undertaken.

Notes

1. Karl Marx, *Capital,* Vol. 1, Modern Library edition, p. 406, n. 2. All citations to the first volume of *Capital* are to this edition. The pagination is the same as in the Kerr edition.
2. This view goes back more than 50 years in the professional economics literature. See Alvin Hansen, "The Technological Interpretation of History," *Quarterly Journal of Economics,* November 1921, pp. 72-83.
3. Karl Marx, *The Poverty of Philosophy* (Moscow, n.d.), p. 105. The sentences which precede the one quoted above make Marx's meaning perfectly clear and reasonable. "Social relations are closely bound up with productive forces. In acquiring new productive forces men change their mode of production; and in changing their mode of production, in changing the way of earning their living, they change all their social relations." Moreover, as Marx points out later, "The handmill presupposes a different division of labor from the steammill" (ibid., p. 127). Surely one need not be a technological determinist to subscribe to these observations
4. Karl Marx and Friedrich Engels, *The Manifesto of the Communist Party*, in Karl Marx and Friedrich Engels, *Selected Works,* Vol. 1 (Moscow, 1951), p. 34. See also *Capital*, p. 823; *The Poverty of Philosophy,* pp. 129-33; and Karl Marx, *Grundrisse* (New York, 1973), pp. 505-11.
5. Marx, *Capital,* pp. 337-38, 367.
6. "While simple co-operation leaves the mode of working by the individual for the most part unchanged, manufacture

thoroughly revolutionizes it, and seizes labor power by its very roots. It converts the laborer into a crippled monstrosity, by forcing his detail dexterity at the expense of a world of productive capabilities and instincts; just as in the States of La Plata they butcher a whole beast for the sake of his hide or his tallow" (ibid., p. 396).

7. Ibid., p. 353. Note also that Marx's panegyrics on the technological dynamism of capitalism apply, *not* to capitalism throughout its history, but only to capitalism as it existed in the century or so before the writing of the *Communist Manifesto*. "The bourgeoisie, *during its rule of scarce one hundred years,* has created more massive and more colossal productive forces than have all preceding generations together" (Marx and Engels, *Manifesto,* p. 37, emphasis added).

8. For a forthright statement of technological determinism, see the work of the anthropologist Leslie A. White. According to White, a social system is "a function of a technological system." Furthermore, "Technology is the independent variable, the social system the dependent variable. Social systems are therefore determined by systems of technology; as the latter change, so do the former" (*The Science of Culture* [New York, 1971], p. 365).

9. Marx, *Grundrisse,* p. 278. Emphasis Marx's. Elsewhere he states: "Whenever a certain stage of maturity has been reached, the specific historical form is discarded and makes way for a higher one. The moment of arrival of such a crisis is disclosed by the depth and breadth attained by the contradictions and antagonisms between the distribution relations, and thus the specific historical form of their corresponding production relations, on the one hand, and the productive forces, the production powers and the development of their agencies, on the other hand. A conflict then ensues between the material development of production and its social form" (Karl Marx, *Capital,* Vol. 3 [Moscow, 1959], p. 861).

10. Marx, *Capital,* Vol. 1, p. 199.
11. Ibid., p. 198.
12. Ibid.
13. Ibid., p. 200, emphasis added. The paleontological mode of reasoning is continued after the passage quoted. Marx adds in a footnote on the same page: "However little our written histories up to this time notice the development of material production, which is the basis of all social life, and therefore

of all real history, yet prehistoric times have been classified in accordance with the results, not of so-called historical, but of materialistic investigations. These periods have been divided, to correspond with the materials from which their implements and weapons are made, viz., into the stone, the bronze, and the iron ages."

14. Simon Kuznets, *Modern Economic Growth* (New Haven, 1966), p. 9.
15. *Capital,* Vol. 1, pp. 371-72.
16. Ibid., p. 504.
17. Ibid., p. 532. For further discussion of these and related issues, see Nathan Rosenberg, "Karl Marx on the Economic Role of Science," *Journal of Political Economy,* July-August 1974, pp. 713-28.
18. For some tentative exploration of these relationships, see Nathan Rosenberg, "Science, Invention and Economic Growth," *Economic Journal,* March 1974, pp. 90-108.
19. "As inventions increased in number, and the demand for the newly discovered machines grew larger, the machine-making industry split up, more and more, into numerous independent branches, and division of labor in these manufactures more and more developed. Here, then, we see in Manufacturing the immediate technical foundation of Modern Industry. Manufacturing produced the machinery, by means of which Modern Industry abolished the handicraft and manufacturing systems in those spheres of production that it first seized upon" (*Capital,* Vol. 1, p. 417).
20. Ibid., p. 418.
21. Ibid., p. 420. See also pp. 417-18 and 421.
22. Ibid., p. 492.
23. Karl Marx, *Capital,* Vol. 3, p. 79.
24. Ibid., p. 100. Later Marx adds: "The most striking example of utilizing waste is furnished by the chemical industry. It utilizes not only its own waste, for which it finds new uses, but also that of many other industries. For instance, it converts the formerly almost useless gas-tar into aniline dyes, alizarin, and, more recently, even into drugs" (ibid., p. 102).
25. Ibid., p. 80.
26. Ibid., p. 81. See also p. 84.
27. Ibid., pp. 81-82.
28. Ibid., p. 71.
29. "From the moment that the tool is taken from man, and

fitted into a mechanism, a machine takes the place of a mere implement" (*Capital*, Vol. 1, p. 408).
30. Ibid.
31. Ibid., p. 420.
32. Ibid., p. 418.
33. See Abbott Payson Usher, *A History of Mechanical Inventions*, rev. ed. (Cambridge, Mass., 1954), especially Chapter 4, "The Emergence of Novelty in Thought and Action."
34. In all of this, the natural environment plays a critical role. "It is not the tropics with their luxuriant vegetation, but the temperate zone that is the mother country of capital. It is not the mere fertility of the soil, but the differentiation of the soil, the variety of its natural products, the changes of the seasons, which form the physical basis for the social division of labor, and which, by changes in the natural surroundings, spur man on to the multiplication of his wants, his capabilities, his means and modes of labor. It is the necessity of bringing a natural force under the control of society, of economizing, of appropriating or subduing it on a large scale by the work of man's hand, that first plays the decisive part in the history of industry" (*Capital*, pp. 563-64).
35. *A History of Mechanical Inventions*, p. 78.
36. Ibid., p. 65.
37. *Capital*, Vol. 3, p. 112.
38. Ibid.
39. Ibid., p. 103, and *Capital*, Vol. 1, p. 442. Marx cites Babbage for supporting evidence in both places and Ure in the former. Both arguments—with respect to improvements in machine design and improvements in techniques for producing the machines—are combined in *Capital*, Vol. 1, p. 442: "When machinery is first introduced into an industry, new methods of reproducing it more cheaply follow blow upon blow, and so do improvements, that not only affect individual parts and details of the machine, but its entire build. It is, therefore, in the early days of the life of machinery that this special incentive to the prolongation of the working day makes itself felt most acutely."
40. See Nathan Rosenberg, "Factors Affecting the Diffusion of Technology," *Explorations in Economic History*, Fall 1972, pp. 3-33.

Social Relations of Production and Consumption in the Human Service Occupations
by Gelvin Stevenson

The human services require an explicit theoretical analysis of the social relations of consumption in addition to the social relations of production. The purpose of this paper is to begin such an analysis.

Certain definitions are in order at the outset. Service industries, broadly defined, include: retail and wholesale trade, finance and insurance, real estate, households and institutions, general government (including the military), and professional, personal business, and repair services. The human services is a narrower term, applying only to those industries of which the primary outcome is an interaction designed to change the characteristics or condition of one of the people involved in the interaction, i.e., the consumer. The human services industries include education, health (mental as well as physical), social service, and to some extent police services. They exclude retail trade and other industries where provider and consumer interact, but where the intent of the interaction is a sale, not a change in the physical, mental, or emotional state of the consumer.

History

The development of monopoly capitalism over the past century has increased the contradictions of capitalism, heightening tensions and antagonisms between classes and extending the hegemony of market relationships over human relationships. Increased capital accumulation and concentration effected both

I am indebted to Charles Valentine for many useful comments on this paper.

the demand for and the supply of human services. First, the need for new forms of social and health services as well as for social control grew due to the increased proletarianization and urbanization of the population, the increased numbers of elderly and dependent people, and a general increase in alienation. In other words, there was an increasingly individualized, atomized populace containing rapidly rising numbers of displaced persons. Second, the social institutions which had previously provided the services that people need as well as the social control required to maintain the status quo were being destroyed: nuclear and extended families became smaller and more dispersed, due to economic and consumptionist pressures and increased urbanization and geographic mobility; churches fell victim to increased secularization; and the sense of community in urban areas was destroyed by "renewal," racism and ethnic intolerance, market pressures, and other heightened antagonistic relationships characteristic of capitalism.

These trends have triggered an important development in the activities of the state in monopoly capitalism. Originally arising out of the need to keep class antagonisms in check, the state in monopoly capitalism has had, among other things, to expand its activities in order to satisfy two sets of demands: those of the ruling class to keep the expanded and heightened antagonisms in check; and those of the working classes to provide needed public services, such as education, medicine, and protection. The state, in response to these pressures, has developed subtle and effective ways of quieting class antagonisms while appearing to provide, and in some cases actually providing, services. The state, in order to protect itself and the dominant classes, not merely hires more and more policemen to control the working and lumpen classes, although it definitely does do that. The state has also taken to financing or providing "human services." These activities—primarily schooling, medical care, social work, rehabilitation, and protection—provide, simultaneously, elements of service and control. In this way the state satisfies both sets of demands on it—services demanded by working and poor people, and control demanded by the ruling classes.

The tremendous demands for services and controls caused

rapid and extensive growth in these industries. From 1960 to 1970 the number of workers in these industries increased enormously, as shown in Table 1. Together, employment in

TABLE 1
*Employment Growth in
Human Service and Other Industries
1960 to 1970*

Industry	1960	1970	Percent Change
Education	3,385,207	6,138,324	81.3
Health	2,589,253	4,247,671	64.1
Hospitals	1,691,578	2,680,066	58.4
Other health	897,675	1,587,605	76.9
Welfare and Religious services	608,581	856,526[a]	40.7
Total	9,172,294	15,510,192	69.1
Manufacturing	17,529,762	19,864,209	13.3
Total, all industries	64,646,563	76,805,171	18.8

[a] Includes religious organizations, welfare services and residential welfare facilities.
Source: U.S. Department of Commerce. Bureau of the Census. *U.S. Census of Population: 1960. Industrial Characteristics.* Table 2. *U.S. Census of Population: 1970. Industrial Characteristics.* Table 32.

these industries increased 69.1 percent compared to an 18.8 percent growth for total employment and a 13.3 percent growth in manufacturing employment. The same trend was operative in other capitalist countries.

During the sixties, members of the working and lumpen classes struggled over the content of these services: whether they were going to be more service-oriented or more control-oriented, and their availability and distribution. Oppressed people struggled in this arena over, for instance, the control of public schools, welfare rights, day care, health care, and police protection and brutality.

The success of these struggles was sadly limited. Struggles over control were defeated or channeled into economistic directions (more teachers, more doctors, more social workers) that left untouched the basic economic and social structures which had originally created the need for struggle. The outcome has been an "unstable equilibrium of compromise" which, according to Gough, "provides the basis for the whole series of social and economic reforms extracted by the working class in the postwar 'welfare states' of advanced capitalist societies, which yet leaves untouched the political power of capital and the repressive apparatus of the state on which it is ultimately based."[1]

Because they were mediated through the existing socioeconomic system, and partly because those struggling for change did not fully understand the repressive potential of human services, the reforms which were instituted as a result of these struggles served "simultaneously to provide tangible benefits to various elites and symbolic benefits to mass publics, quieting potential unrest, deflecting potential demands, and blurring the true allocation of rewards."[2]

It appears that in the immediate capitalist future, deepening economic crises will lead to increased structural unemployment, more diseases of civilization (diabetes, cancer, stress, etc.), and heightened social antagonisms. These in turn will cause a continued increase in the numbers of students, unemployed, patients, counselees, and other consumers of services. And this phenomenon will tend to increase the numbers and the significance of human service workers. Also, the impacts which these workers have on the political consciousness of members of advanced capitalist societies will continue to increase. Their roles of serving and controlling the mass of people make them touchstones of the system—the places where many people first come in contact with the organized system. For the lumpen class, whose members do not experience the workplace for sustained periods, they serve as at least a partial replacement of the workplace as a fountainhead of consciousness. Human service workers therefore will continue to grow in numbers and importance. This is a particularly significant trend because, as we shall now see, they have a unique and expanding role to play in either radicalizing or oppressing consumers.

The Nature of Human Service Industries

Human service industries differ in several important ways from manufacturing and extractive industries. First, in human service industries production and consumption occur simultaneously. They are part of the same act. The product is nothing material, but rather a form of interaction between two or more people and a change in one or both of them. It is intangible. Therefore nothing is produced unless it is also consumed. A teacher "produces" teaching when and only when a student "consumes" that teaching. This is in sharp contrast to manufacturing or extractive industries in which the objects of consumption and production are material, concrete objects.

Consumption and production of material goods are dialectically united and neither could exist without the other. Production creates the material object for consumption, and consumption realizes or actualizes production. In this respect, service and other industries are similar. But, since the product of non-service industries has a concrete material form and that of the former does not, it is only in the non-service industries that "consumption completes the product as a product by destroying it, by consuming its independent form."[3]

Second, it follows that in the human service industries, the social relations of consumption are much more closely and directly related to the social relations of production. In a school, for instance, teachers, like workers in other industries, enter into certain social relations of production with their peers—other teachers—and into quite different social relations of production with those above them in the hierarchy—administrators for teachers, owners or their representatives for manufacturing or extractive workers. But only for human service workers do there also exist social relations with the human object of their labors. Only with service workers are there direct and immediate social relations with the consumer.

These social relations comprise a component of the social relations of production that does not exist in other industries, i.e., between the "worker" and the consumer (or object of the work). These social relations also comprise the social relations of consumption. Again, a comparison with manufacturing industries will be useful. There, relations of consumption occur spatially and temporally removed from those of production, although they reflect and are ultimately determined by relationships within the sphere of production. Relations of consumption, such as status, hierarchy, and relative comfort and privilege are determined or at least partly determined by commodities, i.e., by items of consumption. There is a definite relationship implied between the drivers of a high-priced Corvette and of a low-priced Nova.

In human service industries, these social relations of consumption are generated at the point of production. Feelings of status, hierarchy, self worth, etc., are generated in the interaction between the provider and consumer, for example by whether one visits an upper-class physician (Park Avenue) or a lower-class physician (Medicaid mill). There are very different status associations, and these are reinforced by the treatment received at the hands of the providers.*

There is another difference. In the case of material goods, relations of consumption are more flexible. A worker or member of the petty bourgeoisie may buy a Corvette and thereby change his relation of consumption. He or she will then be identified as someone who owns a Corvette (although generally not by other Corvette owners who would be aware that this interloper lacks the true vestiges of the ruling class, i.e., other expensive consumptionist items). But in human service industries it is much more difficult to buy or suddenly acquire a change in class identification, because the service worker continually reinforces one's class identification. The working-class person who goes to the Park Avenue physician will be made to feel out of place. And this will occur at the point of production/consumption where the provider and consumer interact. This leads us to our third point.

Third, service workers play a direct and active role, at the point of production, in the social control of the consumers of their services. This takes several forms. In order to receive

---

* Yet while we focus on the differences between types of industries, we must not lose sight of the similarities. For in manufacturing too, the social relations of consumption—all types of social relations in fact—are based ultimately on social relations of production. In a class society, one's status as worker or capitalist generally determines what and how one consumes and to whom and how one relates personally.

the services offered they must—at least in a capitalist society—enter into a subservient relationship with the service worker—the "provider." (This is most valid when the provider has a higher class status than the consumer, less so when the consumer comes from the higher class.) They must become a "client" to a "professional." Or a student to a teacher, a civilian to a policeman, a patient to a nurse or doctor, a case to a caseworker.

In other words, here more directly than anywhere else, "Production . . . produces . . . the mode of consumption."[4] Since all cars are produced for conspicuous consumption, using a car necessarily requires a certain amount of conspicuousness. Similarly, when a teacher lectures, a student can consume only by listening. When a social worker "talks down" to a client, the client either accepts that mode of consumption and reacts in a servile manner, or resists and either refuses the service or intimidates the social worker into altering the production/consumption mode. In the hospital, the physical plant, in the form of room accommodations, determines how one consumes medical treatment. In at least one case, it has been reported that "the physical plant was constructed to enable the social system of the community to function . . . without too much distortion."[5] And, as Marx put it, "the object is not simply an object in general, but a particular object which must be consumed in a particular way, a way determined by production."[6]

Production also creates the need for consumption. "The need felt for the object is induced by the perception of the object."[7] In material industries, this is obvious, although the advertising—the mass production of felt needs—that accompanies material production has subtle components. Nevertheless, there is no general desire for white-wall tires until they are produced. In human service industries, the creation of needs and desires for consumption are more subtle and insidious. On one level, professionals tell clients what they should consume. The need is created directly. Dentists tell us to have our teeth checked at least once every six months. A physician tells us to see several different specialists and then return to him. On another level, human service professionals make clients dependent on them. Psychiatrists and social workers may make their clients psychologically dependent on them. Physicians may make their patients physically dependent, by being the sole dispenser of dependency-inducing (physically addictive or not) drugs. Professionals treating drug addicts make their clients physically dependent on another drug (methadone or antabuse, to block desires for heroin or alcohol, respectively) or, perhaps unintentionally, psychologically dependent on a particular way of dealing with authority (complete acceptance) and peers (confrontation). Teachers create dependency through the use of grades and presenting themselves as possessors and dispensers of knowledge.

To sum up these two points: "Production thus produces not only the object of consumption but also the mode of consumption, not only objectively but also subjectively. Production therefore creates the consumer."[8]

Contradictions in the Service Relationship

The above discussed characteristics of human service industries and the social relations that occur therein give rise to several inherent contradictions. A brief discussion of these contradictions should help clarify the dynamics of these industries. It should also show why debating the role of human service workers as *either* serving *or* controlling is static, ahistorical, and misleading. The point here is not that social control *per se* is evil. Every human society requires some form of social control. The political issues revolve around the questions of the goals and methods of social control and the process by which they are developed and carried out. Presumably this contradiction is an antagonistic one under class societies but would become non-antagonistic in a classless society.

(1) *Service Work as Containing Social Service and Social Control.* By examining the point of production—which in this case is also analyzing the commodity—we see that two opposite tendencies are provided in the service relationship—social service and social control. As in any contradiction, one or the other of these aspects may be dominant in any particular situation. For example, the service aspect is dominant in New York City schools when teachers teach white middle-class students. The

control aspect is dominant when they teach black and Latin working-class students. Thus the reading achievement curve in New York's public schools is bimodal. And in any particular situation the more a student objects the more the control aspect is made to dominate. In many cases these two aspects are presented as part of the same act. Physicians tell patients to follow orders if they want to be cured. In other words, submit to the social control as a condition of receiving the social service.

(2) *Service Workers as Exploited and Oppressors.* We now see how service workers are employed as oppressors via their exercise of social control. In their role as workers in capitalist society and as a result of their production process they are also exploited. Therefore a service worker embodies a contradiction in that he or she is both oppressed and an oppressor.

An analogy between providers and U.S. soldiers in Vietnam may help make this point. U.S. soldiers were oppressed by being sent to kill and possibly be killed. They were also oppressors in a very direct and immediate way. Likewise teachers are exploited when they are sent into overcrowded classrooms with children who desire to express themselves individually or in small collectives—but noisily in either case. They are also oppressors when they don't teach anything or do communicate racist, sexist, class-biased, and ethnic-biased views or ideologies. The same contradiction exists in all human service workers.*

One important implication of this contradiction is that certain struggles of progressive elements among human service workers, i.e., traditional bread-and-butter issues like higher salaries and better working conditions as well as more fundamental struggles around democratization and control, may not have any bearing on their role as oppressors. The (New York City) United Federation of Teachers is a good example of a union that has effectively fought against the exploitation of its

---

* It also exists for other workers, although in a somewhat different way. Factory workers are oppressed and exploited. But they also make a necessary contribution to the oppressive control functions of industry insofar as they cooperate with the established control and production processes. One implication is that social change will not have a truly humane outcome until most workers in most industries actively oppose the oppression within and created by their particular institutions.

members while increasing their oppression of the students, their consumers.

Progressive human service workers must deal explicitly with the amount of control and service they are providing and how they relate to consumers. They must evaluate their roles as providers of control (oppressors), providers of service, and consciousness-raisers.

A cautionary note is due the reader at this point. This analysis is meant to develop categories and concepts. It does not include a discussion of the historical development of material production or of the human services. Thus it is in an important sense ahistorical, while a comprehensive analysis must be historical. A crucial element of that history is the evolutionary nature of the distancing between the social relations of production and consumption under different modes of production as well as within the capitalist mode. For example, in pre-industrial modes, the social relations of production and consumption were more closely united than under capitalism. Further, trends in the capitalist human services industries, e.g., teaching machines and computerized medical diagnoses, may increase the distance between the relations of production and consumption in these industries.

Notes

1. Kathleen Gough, "State Expenditures and Capital," *New Left Review* 92, p. 65.
2. Robert R. Alford, *Health Care Politics: Ideological and Interest Group Barriers to Reform* (Chicago, 1975), p. x.
3. Karl Marx, *A Contribution to the Critique of Political Economy* (Moscow, 1970), p. 198.
4. Ibid., p. 197.
5. Duff and Hollingshead, "The Organization of Hospital Care," in Hanz Peter Dreitzel, *The Social Organization of Health* (New York, 1971), p. 234.
6. Marx, *A Contribution,* p. 197.
7. Ibid.
8. Ibid.

The Other Side of the Paycheck: Monopoly Capital and the Structure of Consumption

by Batya Weinbaum and Amy Bridges

I

The housewife is central to understanding women's position in capitalist societies. Marxists expected that the expropriation of production from the household would radically diminish its social importance.[1] In the face of the household's continuing importance, Marxists have tried to understand it by applying concepts developed in the study of production.[2] Yet obviously, the household is not like a factory, nor are housewives organized in the same way as wage laborers.

As Eli Zaretsky has written, the housewife and the proletarian are the characteristic adults of advanced capitalist societies.[3] Moreover, households and corporations are its characteristic economic organizations. Just as the socialization of production has not abolished the housewife, so accumulation has not abolished the economic functions of the household. Harry Braverman has demonstrated how the accumulation process creates new occupational structures, and he has documented the expansion of capital's activity to new sectors. We will argue that these developments also change the social relations of consumption, an economic function which continues to be structured through the household and performed by women as housewives.

We will show how capital organizes consumption work for housewives, drawing them out of the household and into the

---

Many people read earlier drafts of this article and provided important criticisms and insights. Our discussions with friends have furthered our understanding, and we would like to thank Carol Benglesdorf, Carol Brown, Maarten deKadt, Rosalind Feldberg, David Gold, Sherry Gorelick, Heidi Hartmann, Ira Katznelson, Paula Manduca, Alice Messing, Laurie Nisonoff, Rosalyn Petchesky, Frances Piven, Adam Przeworski, Ann-Marie market. The changing relations of consumption work require more time to be spent outside the house, and create a context in which housewives develop their own political perspectives on capitalist society. In particular, the context of housewives' political consciousness will be found in the contradictions between their work in the market and their role in the home. We think that this aspect of women's activity provides a perspective for viewing women's work inside the home, and women as wage laborers, about which a great deal has recently been written. We will argue that capital makes contradictory demands on women's energies, structuring conflicts for individual women, and structuring conflicts between housewives and wage laborers in the market. These arguments require an understanding of capitalism in which we can locate consumption, which is the purpose of the next section.

II

In every society, people must have food, clothing, and shelter in order to live. In capitalist society, production of these necessities is organized for private profit, and people must acquire the things they need for survival by buying commodities. Therefore, as capital expropriates production from households, it also expands market relations. These, like production relations, are "definite relations that are indispensable and independent of [our] will."[4] The obvious consequence of monopoly ownership of the means of production is monopoly ownership of commodities, and the necessity of *purchasing* the means of life.

Insofar as capitalist production is reconciled with social needs, this happens in the market. In a society of small, independent producers, sellers brought their products to the market for exchange. Only in the market would they discover if their product filled a social need. Since producers worked indepen-

Traeger, and Nancy Wiegersma for their help. Discussion of this paper at various places it was presented has also been helpful, and we would like to acknowledge the Womens Studies College of SUNY Buffalo, Chicago URPE, Northeast Regional URPE, Boston University Sociology Colloquium, and Marxist Feminist Groups I and II, for providing forums for us to discuss these ideas.

dently, rather than coordinating their activities, the outcome was chancy. If the product was salable, its price (money in the pocket of the producer) placed constraints on the producer's ability to fill his or her needs. So the "social character of each producer's labor" only showed itself "in the act of exchange,"[5] and the market was the place where private production and socially determined needs were—more or less—reconciled.

In advanced capitalist society, the organization of production as a whole retains anarchic characteristics, but large-scale production makes the "social" character of production apparent in the workplace. And "markets" are not organized for individuals to exchange their products. Rather, selling is an activity organized by capital—increasingly, by large-scale capital replacing "Ma and Pa" stores. Yet just as the small producer measured the "social worth" of his product by its price, so wage-laborers largely measure their social worth by the size of their paycheck.[6] And just as the price (small) producers received for their products placed constraints on the ability to meet needs, so income constrains access to commodities. Thus the relation of private production to social needs continues to be evident in the market: consumption via the market is the other side of the paycheck. Just as in all societies people work while in capitalist societies people labor,[7] so in all societies people reproduce themselves, but in capitalist societies they consume. In capitalist societies, the market serves as the bridge between the production of things and the reproduction of people.

The reproduction of people happens in the household. By this we mean simply that the household is the place where people's needs for food, rest, shelter, and so on are met. Of course the household is not a self-sufficient unit containing resources to meet these needs. Household members must enter the labor market to exchange their labor power for wages, and they must also go out to exchange wages for needed goods and services. Most households are made up of families, in which men are the primary wage-earner, and women are responsible for consumption. In the labor market men confront capital in the form of their employers; in the market for goods and services women confront capital in the form of commodities. This sexual division of labor is not absolute: increasing numbers of women work for wages, and many men participate in consumption work.[8] However these roles are divided, household survival requires participation in exchange relations.

Yet the contradiction between private production and social needs remains. Capitalist accumulation creates its own necessities: the reserve army of labor is the clearest expression of capital's needs, which contradict and take precedence over people's needs for their own reproduction. By saying the market is the bridge between private production and social needs, we draw attention to the fact that people must express "effective demand" to get what they need (they must have money). Of course, effective demand is not a matter of choice, for income is determined by position in the class structure. Thus consumption is always a function of class, and when we say that capitalist production is reconciled with social needs, this is always with the recognition that this reconciliation is imperfect under capitalism.

While the market provides the setting for the reconciliation of private production and socially determined need, that reconciliation is primarily the work of women. Women are responsible for "nurturance," and while nurturance requires many kinds of activity, in its concrete aspects it can only be accomplished through the careful management of income. Consumption (purchasing goods and services for household members) is the first step in this task, and it is the housewife's responsibility for nurturance which conditions her confrontation with capital in the form of commodities. Thus the work of consumption, while subject to and structured by capital, embodies those needs—material and non-material—most antagonistic to capitalist production; and the contradiction between private production and socially determined needs is embodied in the activities of the housewife.

III

Consumption is the work of acquiring goods and services. This work is the economic aspect of women's work outside the paid labor force, and we term women doing this work "consumption workers." The term is not meant to imply that women in this role are themselves wage laborers, but it is used to emphasize that what they are doing is work.[9] As already explained, given housewives' responsibility for the home, consumption work is part of the attempt to reconcile production for profit with socially determined needs. In addition, consumption work involves a set of relations between housewives as consumption workers on the one hand and wage laborers in stores and service centers on the other. We will examine consumption work from the point of view of the housewife, and then look at relations between consumption workers and wage laborers in the market.

Ellen Willis was the first leftist to write about "consumerism" as work necessitated by capital, and to insist that understanding "consumerism" as neurotic is simply sexist.[10] Other writers have been more likely to see women as consumers trying to "compensate" for being cut off from socially organized labor by buying things![11] As the means of production have been progressively expropriated from the household, and as capitalists produce commodities which can be more economically bought than made there,[12] the sphere of the market and the necessity for finding things we need there expands. The main impetus to consumption work is not a psychological need to express creativity through purchasing (though keeping a family going on what most people earn is indeed a creative undertaking, with its own gratifications). The force behind consumption work is the need to reconcile consumption needs with the production of commodities.

Housewives' work, therefore, cannot be understood if we see women as simply "sweeping with the same broom in the same kitchen for centuries."[13] And while many men are accustomed to saying that "women are their own boss" and can arrange their work as they will, a careful examination of housewives' work shows that capital and the state set quite a schedule for them. Leaving aside the fact that young children are demanding and insistent taskmasters, the hours of the husband's work, the time the children must be in school, and for households that live from week to week (which is *most* households) the day of the shopping, are not determined by the housewife herself. Housewives must work in relation to schedules developed elsewhere, and these schedules are not coordinated with each other. Housewives are expected to wait for weeks for installations and repairs, to wait in lines, to wait on the phone. Changes in the distribution network and the expansion of services demand physical mobility within this less-than-flexible series of schedules. The increase in the number of services as well as shopping centers means housewives spend more time travelling between centers than in producing goods or services. The centralization of shopping centers and services may make distribution more efficient, but at the expense of the housewife's time.[14] The consumption worker, unlike the wage laborer, has no singular and obvious antagonist, but many antagonists: the state, the supermarket, the landlord, etc.

Examination of consumption work also requires analysis of the division of labor between paid and unpaid workers in shopping centers. Relations of production in these sectors reappear in a corresponding structure of consumption work. Here the consumption worker frequently plays an important part in affecting productivity. Ben Seligman illustrates this mechanism with the example of retail food centers:

> It is sometimes argued that gross margins have gone up since the 1950s because modern supermarkets' methods shift the burden of services to the housewife. No longer is a human clerk available to advise her as to which product represents the superior buy; the clerk has been transformed into the "materials handler," stamping prices on canned goods, and the only information he is able to impart concerns the location of the canned beans. In effect, the housewife now performs services that at one time were paid for by the retailer. In Switzerland an effort has even been made to have supermarket customers punch their own cash registers (it has not met with success). The housewife performs more and more tasks—searching the shelves, grinding the coffee, filling the basket—and contributes to the upward drift of the margins because she is not reimbursed for her services. Of course, she ought to be paid in the form of lower prices, but in the present course of events, that seems unlikely.[15]

The same holds true in retailing, health, education, and other service industries:

> In the supermarket and the laundromat, the consumer actually works, and in the doctor's office the quality of medical history the patient gives may influence significantly the productivity of the doctor. Productivity in banking is affected by whether the clerk or the customer makes out the deposit slip—and whether it is correctly made out or not. Thus the knowledge, experience, honesty, and motivation of the consumer affect service production.[16]

Capital, therefore, demonstrates this ability to increase its own profit by rearranging the labor process and working conditions of shopping and service centers. Those employed there find their work increasingly reduced to detail labor; those who shop for services do the walking, the figuring, the comparing, and sometimes even the services themselves (as when auto drivers fill their own gas tanks). Each center has its own rules of behavior and performance. Both those who are employed and those who are shopping or seeking services suffer a speedup.

As we have indicated, consumption work is not just buying "things," but also buying services. Just as it has become more economical to buy many things than to make them (bread, clothing, chicken soup), so "the care of humans for each other has become institutionalized,"[17] and households have become increasingly dependent on securing services from the state and through the market. The expansion of services has been undertaken both by the state (education, welfare, prisons, old age homes) and by capital (some medical services, some old age homes, insurance, banks, fast-food chains, laundries, hairdressers). Together, and with the absence of reasonable alternatives, they render households increasingly dependent on a proliferation of widespread centers.

This transition is most vividly demonstrated in changes in the organization of medical services. At an earlier stage of capitalism, doctors could carry a bag of tools to make house calls. The doctor who now relies on an array of testing equipment can only provide medical care in hospitals and clinics, and housewives must bring family members to them. Indeed, there, as in other service centers, the housewife is little more than a detail laborer, lacking access to expertise to judge the quality of what she gets, power to choose what she will purchase, or the ability to replace the service with a self-organized counterpart. Even the women's health movement, for example, while it can provide many kinds of routine care, has barely begun to appropriate the expertise of the medical profession and re-work medical science to be more useful to women.

At times, particular developments in the accumulation process draw more women into the paid labor force. At present, the expansion of the service sector and of clerical work,[18] in conjunction with the fall of real wages among men, pushes increasing numbers of women into the labor force. Just as consumption work requires increasing time and energy,[19] fewer women are able to provide that time and energy. While capital enters new arenas of activity, it continues to organize them in an anarchic rather than a socially coordinated way. The needs of capital are contradictory, therefore, in regard to its demands for women's time. Worse, in a recession public funding for services declines, and work we are increasingly ill-equipped to perform is pushed back into the home. Day care centers close;[20] schools go to double sessions (making it harder to coordinate children's school hours with parents' work hours); Mayor Daley even encourages neighborhood vegetable gardens! Since women are usually both the consumption workers and the wage laborers in the distribution of goods and services, it is especially clear that capital shifts between paying and not paying for the same work. The wage laborers in the commercial and service sectors have strikebreakers perpetually at their door. Capitalist organization pits cashiers and shoppers, nurses and patients, teachers and parents, against each other.

There are, of course, class differences in the work of "housewives." Ruling-class women need not concern themselves directly with reproduction on a daily basis, though they do have a particular role in the reproduction of capitalist class relations. Charity activities, for example, smooth the rough edges of capitalism and help legitimate the social system as a whole.[21] Our sketch of housewives' work is most representative, we believe, for working-class and so-called middle-class women. We may, however, make some distinctions between them. More income gives middle-class women freedom from the more degrading aspects of consumption work (they can have their groceries delivered). These women may also make consumption a "creative" activity, and a means of self-expression. This is no doubt the basis for the idea that *all* women engage in consumption for its psychological benefits. Finally, middle-class women take upon themselves the responsibility for organizing others' consumption, through voluntary organizations.[22] Working-class housewives more often participate in the wage-labor force, thereby taking on a second job. Lower income makes consumption a complex survival task. Women who are dependent on the state for support obviously spend more time obtaining both goods and services from civil bureaucracies than other women do; in addition, the commodities available to them are overpriced and of poor quality.[23] Thus, capital constructs consumption work for women in complex ways: capital organizes the distribution of income to the household, and this largely determines the distribution of households into neighborhoods; at the same time, capital organizes distribution of particular goods and services to particular areas.

We have argued that consumption work is structured by the state and by capital, and that this work is alienating and exhausting. The reproduction of labor in *capitalist* societies requires that the products and services produced with a view to profit be gathered and transformed so that they may meet socially determined needs. In this situation, it is not clear what kinds of reorganization will take place. Certainly, ideas for the reorganization of consumption work on a social basis have been around for a long time (cf. Gilman's *Women and Economics*). Yet the reorganization of consumption work and services to living labor on the part of capital and/or the state can hardly be expected to result in humanized social services. The experiences of and proposals for state-run child care are a case in point, that the profusion of goods and services under capitalism results in increased dehumanization.

There is nothing in shopping, or going for health care or education *per se* that must be alienating and tiring. After all, for centuries the market was the site of social interaction, and a time for holiday. It is housewives' responsibility for "nurturance" on one hand, and the impossibilities of helping other human beings be healthy and creative within the constraints of the present system on the other, that create the incredible tensions of the practice of consumption work.[24] As Roz Petchesky says:

> It's the connection between the shit private production provides in the market and the miracles women are supposed to perform with it inside the family that's really the key. The cutting edge of consumption work isn't procuring but taking up the slack—trying to maintain goods designed for obsolescence; trying to prepare nourishing meals out of vitamin-depleted, over-processed foods . . . trying to encourage and tutor kids that the schools doom to failure.[25]

For all her efforts, the housewife lacks the social power to provide what she feels is best for her family. It is consumption work on one hand, and the ends which it is supposed to serve on the other, which form the network from which housewives' perspective on society is developed.

IV

How has this perspective been organized in practice?[26] In the first instance, consumption work leads to specific areas of political activity, for example, housing. As explained by an organizer in a Boston tenants' union: "The majority of workers in the tenant movement are women. An explanation for this is that tenants' unions are an area where women can be aggressive and take on an active leadership role because we are spending a great deal of time where we live and know the people we live with."[27] Similarly, boycott activities, militant responses to inflation (especially of food prices), and community struggles (often directed against state policies), are areas in which women play important, if not predominant, roles.[28]

But more generally, the dispersed organization of consumption workers, prey to many capitalists as well as to the state, seems conducive to recognition of the oppressiveness and exploitation of capitalism as a system. During the Brookside miners' strike, the miners' wives not only supported the demands of their husbands, but also made more radical and far-reaching demands, insisting on food stamps, boycotting and picketing stores, protesting anti-strike propaganda, and harassment of strikers' children in the schools. Their practice as housewives demonstrated to them that not just the workplace, but the whole city was dominated by the mine owners, and their political activity demonstrated this to the community.[29] In cities where the ruling class is more immediately diverse, this perspective is more complicated, but it still underlies many of women's non-workplace struggles.

Women's activity in revolutionary times may flow from activities ordinarily engaged in, but which take on more political meaning during political upheavals.[30] In Portugal since the overthrow of the fascist regime, women in working-class neighborhoods have formed tenant committees to take over buildings for dwelling units and for community service facilities. These tenant committees have survived their initial activities and remain a basic organizational form in urban communities.[31] Similarly, Chilean women were active in the construction of distribution networks before the coup in Chile. During the *Unidad Popular* government, one of the most severe problems was shortages, creating difficulties in food distribution. These problems were in part engineered by rebellious small merchants threatened by socialism, and in part by cattle-growers who slaughtered their herds rather than relinquish them to expropriating cooperatives. These induced shortages and distribution difficulties led to the formation of *Juntas de Abastecimientos* (JAPs) or Prices and Supplies Committees, which were a spontaneous popular response and succeeded in reducing the need for rationing. Housewives played a dominant role in neighborhood groups representing both mass organizations and local retailers. Their task was to ensure fair distribution of consumer goods. In the first month of their existence, 450 JAPs were formed in Santiago, Chile's major city. The committees incorporated 100,000 households and over 600,000 people. Within a few months, 20 percent of the country's beef consumption was distributed through the committees.[32]

Marxists have been too hasty to see community-based struggles as reformist. A struggle is not necessarily progressive because it is in a factory, or reformist because it is outside it.

If leftists have, until recently, been indifferent to community and consumerist politics, this is in part for a good reason: however progressive these struggles may be as agitational or educational activity, ultimately struggles outside production cannot *alone* constitute a revolutionary strategy. And many community-based struggles have *not* been progressive. Yet to ignore these struggles altogether is unfortunate for several reasons. Demands for control, while they may be accommodated, threaten bourgeois hegemony and serve as a practice in self-management, an important component in the socialist alternative. At the same time, they have a positive education function in demonstrating the possibilities of organized action, and revealing the constraints on political activity within capitalism. Moreover, community and household-based demands insist that production and provision of services be oriented to social needs, and in this way embody values antithetical to capitalist production. They call attention to this society's inability to provide for its people. These demands also embody values upon which a socialist society must be built, that society be organized to meet social needs.[33] Finally, as in the case of the Brookside women, housewives' political activity may come from the recognition that not idiosyncratic malfunction, but the organization of society as a whole, is antagonistic to their needs and interests.

A capitalist society creates many social places from which to view capital: places in production and services (machinist, social worker), places in communities (housewife), places isolated from communities (Wall Street). It follows from the nature of capitalist societies that individuals in many *different* social places may discover that society is not organized for them, but against them. Clearly, there are no places whose occupants are automatically revolutionaries. One of our tasks as Marxists is to investigate the perspectives on capitalist societies which are provided by these different social places. We can only do this if our understanding embraces not only capitalist production itself, but also recognizes how capitalist production shapes society as a whole, and shapes the practices of people in particular places as well. We have shown some of the ways capital structures consumption work, organizing the daily practice of housewives, on which their understanding of society is based.[34] The organization of a revolutionary class requires the joining of those perspectives antagonistic to capital, and forging a vision of society collectively organized to meet social needs.

Notes

1. Marx, Engels, Lenin, and Bebel, for example, recognized that women were oppressed in the family. They thought women's liberation and the possibility of healthy relations between men and women would result when the family ceased to be the basic economic unit of society. Within capitalism, men and women would become wage laborers, as the production responsibilities of the household became socialized. With the abolition of private property, services could be socialized as well, and men and women would be free to form personal relations free of economic functions.
2. In their emphasis on work done inside the household, and understood as "production," most Marxist-feminist work could be included: Paddy Quick, Peggy Morton, Mariarosa Dalla Costa, Margaret Benston, Juliet Mitchell, etc. We recognize that housewives prepare goods for use by family members, but our emphasis is not on housework as a kind of "production." Rather, we argue that housewives' activity is largely a reflection of the fact that *capital* organizes the manufacture of goods and provision of services.
3. Eli Zaretsky, "Capitalism, the Family, and Personal Life," *Socialist Revolution,* Vol. III: 1-2, 3, nos. 13-14, 15 (1973).
4. Karl Marx, Preface to *A Contribution to the Critique of Political Economy* in Karl Marx and Friedrich Engels, *Selected Works in One Volume* (New York, 1968), p. 182.
5. Karl Marx, *Capital,* Vol. I (New York, 1967), pp. 107-108. Cf. also the *Grundrisse,* trans. Martin Nicolaus (New York, 1973), p. 225. Marx discusses the market in Chapters 2 and 3 of *Capital,* and in various places in the *Grundrisse.*
6. See *Grundrisse,* trans. David McClellan (New York, 1971), p. 66: "In capitalist societies, the individual's power over society and his association with it is carried in his pocket." For money as a measure of social worth, see *Capital,* Vol. 1, p. 133.
7. Cf. Engels' distinction between work and labor, *Capital,* Vol.

1, p. 186, n. 1. Here Engels writes that the labor process has two aspects: "... in the simple labor-process, the process of producing use-values, it is *work;* in the process of creation of value it is *labor.*"
8. We are not making an argument here about the relation of capital to the sexual division of labor. See Heidi Hartmann and Amy Bridges, "The Unhappy Marriage of Marxism and Feminism: Towards a More Progressive Union," unpublished manuscript.
9. See n. 7.
10. Ellen Willis, " 'Consumerism' and Women," *Notes from the Third Year,* reprinted in V. Gornick and Barbara K. Moran, *Woman in Sexist Society* (New York, 1972), pp. 658-665.
11. Mariarosa Dalla Costa, *The Power of Women and the Subversion of the Community* (pamphlet published by Falling Wall Press), p. 43.
12. See Braverman, *Labor and Monopoly Capital,* p. 281, and Chapters 13 and 16, *passim.*
13. Dalla Costa, *The Power of Women,* p. 36.
14. Centralization of service distribution is economical for capital and the state, but not the best way to provide services, since the services become less accessible. So, for example, when The Woodlawn Organization drew up a plan for Woodlawn Model Cities, an important element was the proposal for neighborhood service centers which would be accessible and would distribute *all* services. Only hospital facilities would be centrally located.
15. B. Seligman, "The Higher Cost of Eating," in *Economics of Dissent* (New York, 1968), p. 229.
16. Victor Fuchs, *The Service Economy* (New York, 1968).
17. Braverman, *Labor and Monopoly Capital,* p. 279.
18. Ibid., Chapters 15 and 16.
19. Capital also increases time spent on women's work done inside the house. See H. Hartmann, *Capitalism and Women's Work in the Home: 1900-1930,* unpublished dissertation, Yale, 1974. See also Walker, "Homemaking Still Takes Time," *Journal of Home Economics,* no. 61 (1969), pp. 621-22.
20. R. Petchesky and K. Ellis, "Children of the Corporate Dream," *Socialist Revolution,* no. 12 (November-December 1972).
21. See G. William Domhoff, *The Higher Circles* (New York, 1971), Chapter 2: "The Feminine Half of the Upper Class."

22. See, for example, Robert S. and Helen M. Lynd, *Middletown* (Cambridge, Mass., 1959) or John R. Seeley et al., *Crestwood Heights* (Toronto, 1956).
23. Largely because of the neighborhoods in which they live. See David Caplovitz, *The Poor Pay More* (New York, 1963).
24. Although our discussion of consumption work has focused on the monopoly capital stage, at places where monopoly capital penetrates prior economic modes, the tension of changes in consumption is also sharp. For example, the purchase and use of powdered milk marketed by international agribusiness corporations in the third world has led to deaths and/or deformities of infants who would have been better off with their mothers' milk, given impure water supply and the need for natural antibodies in the mothers' lactate system. See *Formula for Malnutrition,* CIC Brief, April 1975, available for 60¢ from the Interfaith Center on Corporate Responsibility, Room 566, 475 Riverside Drive, NYC. In the nineteenth century, the demands of capitalist production had the same effect: women workers in England, unable to go home to nurse their babies, gave the babies "Godfrey's cordial," a narcotic which kept the infants asleep but often killed them. See *Capital,* Vol. 1, p. 395.
25. Personal communication to the authors.
26. There is little room for examples, but see Edith Thomas, *The Women Incendiaries* (London, 1967); Alice Bergman, *Women in Vietnam* (San Francisco, 1974); J. Ann Zammit and Gabriel Palma, eds., *The Chilean Road to Socialism* (Austin, Texas, 1973).
27. Barry Brodsky, "Tenants First: FHA Tenants Organize in Massachusetts," *Radical America,* Vol. 9, no. 2 (1975), p. 41.
28. Women also played important roles in organizations created by (or against) the War on Poverty. See J.D. Greenstone and Paul E. Peterson, *Race and Authority in Urban America* (New York, 1973); F.F. Piven and R. Cloward, *Regulating the Poor* (New York, 1971).
29. Ann Marie Traeger and Weinbaum, unpublished interviews, August 1974.
30. Edith Thomas comments: "The 'political' activity of women, then, appeared first in these various consumers' cooperatives; and this follows tradition. Women are much closer to everyday realities than men are. Feeding the family is a part of their age-old role. The price of bread has been their business

for centuries. Thus, before seeking to involve themselves in truly political activity, they tried to attend to 'the administration of things,' upon which they could act directly. It is from this angle that the most aware women among them thought to have a hold on the social reality. But that was obviously only one aspect of the question" (*The Women Incendiaries*, p. 14).

31. Interview with Pacifica Radio correspondent, May 1975.
32. Zammit and Palma, eds., *The Chilean Road to Socialism*, p. 89.
33. The consequences of leaving consumption organized through the household under socialism even after production has been socialized are discussed in Weinbaum, "The Curious Courtship of Women's Liberation and Socialism: Perspectives on the Chinese Case," in the second special issue on the political economy of women, *Review of Radical Political Economics*, Spring 1976.
34. It should be obvious that we are not arguing that all housewives are politically active, much less revolutionary. Just as wage laborers may feel "inadequate" because their earnings are low or because they are not promoted (see Jonathan Cobb and Richard Sennett, *The Hidden Injuries of Class* [New York, 1972]), so housewives may internalize contradictions which are structural. Our emphasis is counter to the current understandings that housewives by nature of their "place" are conservative. For this view, see Zeitlin's book on Cuba, in the preparation of which he didn't interview women because "everyone knows" that women haven't played a role in revolution. Carl Boggs claimed (unpublished manuscript) that one reason the Italian resistance was not revolutionary after the war was "the presence of housewives"; and Weinstein has said that the winning of women's suffrage was a setback for the socialist party, for which his evidence is a single aldermanic election in Chicago, etc., ad nauseam. See also Albert Szymanski, in *Insurgent Sociologist*, Winter 1976.

Marx versus Smith on the Division of Labor
by Donald D. Weiss

I

The division of labor is a particular form of the differentiation of productive functions; just as the differentiation of productive functions is a particular form of social cooperation. Let us look at this more closely.

Social cooperation occurs whenever people work jointly to effect a particular end.

Differentiation of function occurs when those who are cooperating perform qualitatively distinct tasks.[1] Thus if you and I are both pushing a wagon, we are cooperating, but there is no differentiation of function. If, on the other hand, you push while I drive the horses, there clearly are different functions involved.

Division of labor is a special case of functional differentiation. It occurs whenever the various productive functions are performed in such a way that each person is assigned a task as his or her particular *occupation*. If jobs are rotated, then we have differentiation of function but *not* division of labor. This distinction is crucial. Thus, sometimes it is argued that the division of labor is inevitable, because any society has various tasks that must be accomplished—indeed must be accomplished simultaneously. It is inevitable that one person do one thing while another does another. But, of course, all that this truism proves is that *differentiation of functions* is inevitable.[2] To rebut a common adage: even if it is true that, in a given society, "somebody (in particular) must dispose of the garbage," it does

---

This paper was presented earlier this year in Richmond, Kentucky, at a conference observing the bicentennial of Adam Smith's *The Wealth of Nations*. I would like to thank the Rabinowitz Foundation and the SUNY Research Foundation for their generous support.

not follow, without any ado, that there must be garbagemen, i.e., people who spend their entire working life at this one task. And if, say, *each* of us must spend one week per year on garbage detail, then we do not (insofar) have division of labor properly so called.

In short, we have the division of labor only when we have *specialists*. But, just for this reason, such division seems always to involve some degree of special *expertise* on the part of the practitioners of the respective functions. For if I spend all my working time farming, while you spend all yours fishing, then—assuming we are both able-bodied and sound-minded—I will be better than you at farming and you will be better than I at fishing. The concentrated practice of a function generally does make one at least somewhat better at it than those who have not so concentrated.

A correlation is implied in this between the extent of the division of labor and the *efficiencies of expertise*. And this correlation is, in fact, cited by Adam Smith as the first great advantage of specialization. The greater the degree of specialization, the more "dextrous" do people become at their particular tasks.[3] Indeed, one of the two other advantages also discussed by Smith involves a closely related point, though this connection is not explicitly remarked by him. I refer to the *promotion of inventiveness*.[4] Smith claims that people will generally be better at spotting opportunities for technological improvement in functions with which they have detailed acquaintance—an assertion which surely is plausible. I would only point out that it is a little misleading to treat this advantage, as Smith does, as entirely independent of the first. It would have been theoretically more elegant for him to claim that the division of labor promotes expertise, such expertise having at least the following two aspects: it creates efficiency in each phase of production; and it makes the practitioners of each function more sensitive to the ways in which such efficiency can be even further promoted. (The third advantage of the division of labor discussed by Smith will not concern us here; and so reference to it may be confined to a note.[5])

Now given this connection between specialization and expertise, it is natural to assume that there is, in general, a fairly strict correlation between the degree of material culture attained by a people and the degree to which that people has divided up its labor. It would seem that with increased division of labor, we could produce *better goods,* and could produce *more* of them in less time, than we could if everyone were, as the old saying has it, a jack-of-all-trades and master of none. And this is, of course, just the conclusion that Smith does in fact draw.[6]

But it is to his credit that Smith does not let matters simply go at that. He knows full well that the productive advantages of the division of labor are only one side of the story.[7] A quite negative side must also be considered. For insofar as the division of labor advances, it clearly is necessary for each person to concentrate his interests and to cultivate his talents within a more and more restricted domain. The more the division of labor advances, expanding our collective productive capacities, the more restricted, correspondingly, becomes the productive activity of each individual. But notice, now: it is clear that the skills and abilities developed in each person are primarily a function of the particular sort of work he does. And it is also clear that a person's capacity to assimilate intelligently the contributions of his fellows is a function (in turn) of the skills that he has developed. If, therefore, my primary life's activity restricts the development of my sensibilities to one particular function, *I am to that extent culturally crippled.* And we are thus led to a poignant dilemma: the ever increasing specialization which augments the wealth of nations also has a tendency to make each specialist less and less able to *appropriate* that wealth beyond the boundaries of his ever shrinking bailiwick.

Smith proposes no dramatic escape from this dilemma. On the one hand, there can for him be no question of a historical retreat, a return to times when the division of labor was less pronounced. For despite the fact that the division of labor makes each person more and more restricted relative to the totality of material culture about him, it nonetheless *does* increase the collective wealth—the national product—immensely. As a result people are, in an absolute sense, wealthier than they would be if the division of labor were less fully developed. That is, despite the fact that each person is, relative to the totality of culture about him, poorer than his more primitive predecessors were *relative* to the totality of material culture in which *they* were immersed, each person is nonetheless wealthier in *absolute* terms, i.e., has more of life's conveniences, than those same predecessors. The increase of *absolute* wealth is purchased at high cost, that of a decreased wealth of the individual powers *relative* to the level of culture available, but it is an increase of absolute wealth, nonetheless. A historical retreat is therefore out of the question.[8]

The crippling effect of the division of labor must therefore be treated as a necessary evil, and hence at best as a condition to be ameliorated and compensated for, but not to be eliminated. Smith does suggest a strategy for tempering the negative implications of the division of labor: we should provide at least a minimally decent level of education for all.[9] True enough, people's lives are being made narrower and narrower by the work process *per se*; but their horizons can be broadened in the classroom. There, each person will be guaranteed some training and exposure in areas beyond his occupational thoughts and concerns. He will to that extent be more humanized. It must be admitted that this is only a compensation for, not an overcoming of, the crippling effects of the division of labor. But it is certainly better than no compensation at all.

II

The following question was bound to arise: Might there be a way of counteracting, in more than merely compensatory fashion, the ravages of the division of labor—*without* sacrificing productive efficiency? It was Marx who provided the theoretical framework for understanding how the answer to this question could actually be yes. In discussing Marx's views on the division of labor, I will first restrict myself to the question of industrial labor. This restriction will be dropped in section III where I will discuss the division of labor in general.

It is clear that despite Marx's deep admiration for Smith, he nonetheless considered Smith to be a "political economist" in a peculiar, pejorative sense of the term: one who mistakes the conditions of the existing system of production for the necessary conditions of production in general.[10] According to Marx, Smith had noted a quite real correlation: that between the division of labor and productivity; but Smith had failed to see that this was a correlation that could be expected to hold only under particular historical conditions. These conditions, Marx believed, were changing. And once they had changed sufficiently, a *new* correlation would be established: that between increasing productivity and the *abolition* of the division of labor.

Marx stakes himself to no less a proposition than this: the division of labor characteristic of industrial production is in the process of withering away. It is the inherent tendency of capitalism to work toward the abolition of specialization in the industrial sphere.

Marx reasoned as follows. In the first major stage in the development of capitalist production, that of hand production or "manufacture," there is a tendency toward the extension and intensification of the division of labor. For, wherever we have hand production, we have the circumstance that the practitioner of one craft must master certain, usually quite subtle, *physical movements*, while other craftsmen must master other such movements. As long as industry is based upon the mastery, by human beings, of certain physical-manipulative skills, productivity will clearly be fostered by the improved "dexterity" promoted in each worker by the division of labor. From the point of view of each individual capitalist, therefore, the extension of the division of labor will be desirable. The more refined the division of labor, the more productive one's plant and therefore the greater one's profit. Under these circumstances, Marx would freely admit, the Smithian correlation between the division of labor and productive efficiency clearly holds.

But with the introduction of machine production we have the onset of a striking new tendency. A historical point is reached at which the differences between the skills involved in the various branches of industry start becoming less and less pronounced. As production becomes increasingly automated, the skills required to make product A come increasingly to resemble those required to turn out product B. The reason is that *while the physical movements required to produce A and B must, until the age of automation, be mastered by human hands, insofar as automation does take hold, these physical movements are no longer performed by human hands at all.* They come to be done by machines. Insofar as human labor is still involved in production, it tends to be more and more restricted to a narrow range of *maintenance* functions. Unlike the skilled worker, who dextrously wielded his tools, the factory worker comes more and more to be "an appendage of the machine."[11] Plainly, whatever skills are involved in tending an A-producing machine do not differ from those involved in tending a B-producing machine nearly as much as A-producing skills differed from B-producing skills during the period of hand production.

From the point of view of the individual capitalist, it is automation that now becomes the key to greater efficiency and hence higher profits. Thus, the very same "will to profit" that intensified the division of labor during the period of hand production now drives the system into a qualitatively new phase: one in which the differentiation of skills that defines what we mean by the "division of labor" becomes ever *less* pronounced. Under capitalism, the division of labor is first intensified; but after a certain point it begins, parabola-like, to describe a downward path.

But if automation implies a decline in the industrial division of labor, it also implies, Marx maintains, the radical dehumanization of those who remain trapped in factory work. Insofar as jobs require less and less skill, people come to spend their productive lives involved in monotonous tasks which make no demands upon, and hence cannot possibly engage, their intelligence. And this situation will continue as long as the social relations developed by capitalism prevail.

It is at this juncture that Marx achieves a simple yet profound dialectical insight. He sees that the very process—automated production—which dehumanizes the factory worker under capitalist social relations can, given new social relations, emancipate him. The drudgery of factory work is due to its utter simplicity; and the utter simplicity of this work is rooted, in turn, in the circumstance that human physical labor has become a much less significant component of production. In other words, *just because* industrial capitalism reduces skilled labor to unskilled labor, it must be considered *a tendency to make industrial labor more and more superfluous.*[12]

In short, society as a whole needs to devote less and less of its time to factory work. Eventually, Marx thought, this can have only one result: the notion that an entire class of people *must* spend their lives confined to drudgery seems less and less defensible. It begins to enter people's heads that what little factory work—the maintenance work already referred to—does need to be done, could be socially distributed in such a way that no given person need spend much of his time doing it. If everyone did a short stint of factory work each year, it would be possible for everyone to be free from such work for most of the year. No longer need it be the case that only a privileged sub-group of society—the ruling class—is, in being free *from* the drudgery of industrial production, free *to* develop its creative intelligence. It is now possible for all people to devote themselves to the "higher functions"—a circumstance which is, moreover, not only intrinsically desirable, but *also productively useful*; for an advanced industrial society can be even more efficiently administered if the right to the full development of intelligence has been made universal. In short, for Marx, "the division of mental and material labor" can now, finally, be abolished; and for one very simple reason: "material labor" is becoming increasingly obsolete.

For Marx this means that the functional basis of *class distinctions* is being eroded by capitalist development. The essential distinction between a ruling class and a ruled class is, for Marx, that between a class that monopolizes mental/directional functions, and a class that is confined to the sphere of manual work.[13] Insofar as capitalist development renders such manual work less and less necessary, classes lose their historical point and purpose. We thus arrive at the conclusion that *the growing obsolescence of the industrial division of labor, determined by the growth of automated production under capitalism, is at the same time the key to the establishment of a classless society.*

III

We have so far discovered two basic Marxian propositions concerning the division of labor: (1) that capitalism has an intrinsic tendency to abolish the division of labor within the factory by transforming skilled into unskilled labor; and (2) that this very same process is the basis for overcoming the distinction between mental and manual functions, i.e., between classes.

We can also discern a fundamental difference in approach between Smith and Marx. While the former regards the industrial division of labor in essentially static terms, as an eternal *sine qua non* of high productivity in any advanced economy, the latter regards the development of the industrial division of labor as a *process* culminating in the establishment of a technology which renders that division obsolete. In dialectical terms, for Marx, unlike Smith, the industrial division of labor produces the conditions of its own negation.

The reader has, by now, probably gathered that my sympathies are with Marx. This is not to say that I am insensitive to problems regarding the Marxian conception; but rather to say that I believe these problems can be handled with minimal disturbance to the basic insight.

But it must be admitted that there is at least one very large stumbling block in the way of accepting Marx's orientation toward the division of labor—one which we have not so far considered and which, moreover, concerns a very basic aspect of his thinking on this matter. To appreciate this problem, we must drop the restriction imposed at the start of the preceding section: we must consider not only the industrial division of labor, but the division of labor in general.

The problem is as follows. While Marx does advance cogent reasons for supposing that the industrial division of labor is becoming obsolete, this seems the most that the considerations advanced by him could be taken to prove. And yet the language used by Marx in many of his writings suggests that he takes himself to be arguing that "the division of labor"—in general and without qualification—is in the process of being eliminated. And this, it inevitably is objected, commits Marx to a much bolder, and quite implausible, implication: that the division of labor even within the *non*-industrial sphere—i.e., among society's "mental laborers"—is also being, or could also be, eliminated.

That Marx does speak in unqualified terms about the "abolition of the division of labor" is beyond dispute.[14] It would seem also to be beyond dispute that this conception is implausible even for one who accepts the possibility of the elimination of class distinctions. The literal and complete abolition of specialization would seem to involve transforming everyone into an entirely "universal man," who is learned in all branches of inquiry. It would involve abolishing the circumstances under which we are entitled to say such things as, "Ludwig is a philosopher while Marx is a sociologist." But since each of us has, on the average, but three score years and ten, such a conception seems wild. Given the shortness of life, there is only so much a person can do. A rare individual might, like a latter-day Leonardo, astonish us with mastery of four or five different disciplines. But a literally "universal man"[15]—this would appear to be out of the question. It is nonetheless the implication that seems to be suggested by Marx's own words.

Whoever would try to defend Marx must take one or the other of two courses: argue that such "universality"[16] is not as absurd a prospect as it seems; or else argue that Marx meant something somewhat less radical by the expression, "the abolition of the division of labor," than these words might seem to suggest.

The latter course is the correct one; and the key to the solution of our problem is contained in one of Marx's most famous, yet also most puzzling, pronouncements concerning the division of labor. It is worth quoting at some length:

> [A]s soon as the distribution of labor comes into being, each man has a particular, exclusive sphere of activity, which is *forced upon him* and from which he cannot escape. He is a hunter, a fisherman, a shepherd or a critical critic,[17] and *must remain so* if he does not want to lose his means of livelihood; while in a communist society, where nobody has one exclusive sphere of activity but *each can become accomplished in any branch he wishes*, . . . it [is] possible for me

to do one thing today and another tomorrow, to hunt in the morning, fish in the afternoon, rear cattle in the evening, criticize after dinner, *just as I have a mind,* without ever becoming hunter, fisherman, shepherd, or critic.[18]

I have emphasized certain phrases in this passage in order to indicate that the crucial contrast is that between the *unfreedom* of those in pre-communist societies and the *freedom* of those in communist society. The latter are, while the former are not, free *to do whatever they wish*. Marx speaks here of a freedom "to become accomplished" in whatever field "I have in mind" to become accomplished in. There is no suggestion that each person might literally become accomplished in *all* fields of human inquiry in one short lifetime. The suggestion is, rather, that there will be no restriction on the individual's privilege to work in *any given field* as often as and whenever he pleases.

The solution to our problem thus involves denying that, for Marx, the "abolition of the division of labor" actually means the transforming of each individual into a literally "universal man"; such a conception surely *is* absurd. Rather "communism" consists in the absence of all forms, direct and indirect, of coercion in the sphere of work. *It is the overcoming of the antithesis between what I work at and what I wish to work at.*

Such a conception is bold enough in its own right. We see that, to the two Marxian propositions abstracted at the beginning of this section, we must now add a third: that automated production not only reduces the social need for manual labor to such an extent that the emancipation of the working class becomes possible; it also is responsible for so high a level of productivity that the life-long "fixation" (Marx's term[19]) of each person at one particular task finally becomes socially unnecessary. Whereas, in previous historical epochs, efficient production depended upon the regimentation of functions, i.e., depended on denying to people the right to put down and pick up whatever tasks they pleased whenever they pleased, today, to the contrary, the technology which we have historically developed, and the technological and scientific comprehension which modern culture embodies, make it possible for such freedom to exist without there being any danger of a historical relapse to a qualitatively lower level of productivity.

It is clear that even though this idea is not nearly so wildly radical as the literal "universal man" conception, it may nonetheless seem quite utopian. It will inevitably be objected that it is naive to suppose that such a society would not quickly degenerate into an orgy of dilettantism and sloth. If people are not *made* to do socially useful things, if they are not *forced* to focus their attention on some particular discipline, won't they end up having no socially useful competence at all? And won't this pose the threat that culture's accumulated "technological and scientific comprehension," referred to in the preceding paragraph, will be lost or diminished, and hence that we *will* suffer a "relapse to a qualitatively lower level of productivity"?

Marx does not, to my knowledge, explicitly address himself to these latter questions. This is quite certainly not because they never occurred to him, but rather because he thought them worthy only of disdain. If Marx were with us today, and could be coaxed into making a reply, I think he would speak in something like the following vein:

"There are," he would say, "two basic errors at the bottom of the above objections. It is supposed, in the first place, that the coordination of people's individual plans and activities, essential to the functioning of any society, is incompatible with the conception of freedom I have suggested. It is supposed that, because human affairs would be hopelessly disorganized if society did not have some sort of unified means of planning, it somehow follows that the coercion of the individual by society is somehow inevitable. It is supposed that the claims of society must confront the individual as a denial, as something externally imposed, that a plan of social coordination could not possibly be perceived by people as the fulfillment of their needs. It is supposed, in effect, that there is something about 'human nature' which prevents people from directly desiring the mutual, orderly arrangement of their plans and projects.

"These are just the sorts of assumptions that flourish in the society of the marketplace. Such a society is founded upon antagonistic relations among people, and hence it requires an ideology that apotheosizes just such antagonism. Under capitalism, people confront one another as foes; and if capitalism is to work, such adversary relations must appear normal or inevitable. Thus as long as this system prevails, it is little wonder that we have the prevalence of theories that imply that it is abnormal or impossible for people directly to desire and to seek mutual coordination.

"We communists have never made any secret of the fact that we regard these paradigmatically bourgeois ideas as scientifically defenseless. Let there be no ambiguity about this: the achievement of communism requires the communalization of the human spirit, the creation of needs and desires that are directly cooperative in character, the abolition of the apparently 'necessary connection' between social existence and coercion. And we maintain that there is nothing in 'human nature' to prevent such an achievement. And so we make the following twofold claim: it is true that, in a free society, people will be allowed to pursue whatever life-plans they wish; but *it is also true that each person will wish to make sure that his activities are so ordered that his own enjoyment is at the same time a contribution to the community.*

"But there is a second sort of reason that the 'abolition of the division of labor' may seem to be a utopian conception. This reason is nothing other than the old Smithian assumption that whenever there are many social functions that need to be performed, the radical division of labor is inevitably the only efficient way of performing them. Giving people the opportunity to broaden their productive activities thus appears, from this point of view, as an invitation to dilettantism.

"Now such reasoning was appropriate enough in the period canvassed by Smith; and it remains appropriate for as long as the technological basis of society remains relatively underdeveloped. But in our time we have achieved a fantastic elaboration of technological *matériel*, and a fantastic complexity in human interactions which necessarily go along with it. A point has been reached at which society must begin to place a much higher premium on the development, in each individual, of *a general comprehension of how this complex technical-social whole functions.* We simply cannot efficiently administer such complexity any longer if each person continues to be canalized into just one particular function. People must indeed be allowed to broaden their skills, rather than narrowing them.

"Smith saw that the division of labor implied an unfortunate narrowness in the life of the individual. But what he could not foresee was that, in time, *this narrowness would, beyond certain limits, come to have a deleterious effect upon the production process itself.* What he could not foresee was that the efficiencies of concentration and specialization would someday come to be only one side of the story of human production; that the efficiencies involved in the contrary tendency—that of comprehensiveness of understanding and insight—would someday become of equal significance. Thus if the ethos of the era known and foreseen by Smith rightly stressed the virtues of specific expertise, of *division,* we must, in the coming age, nonetheless reserve an equally hallowed place for the advantages of breadth, of *synthesis.*"

Notes

1. "Differentiation of functions" is my own expression. To my knowledge, it is used by neither Smith nor Marx.
2. The division of labor may of course be "inevitable" for other reasons.
3. Adam Smith, *An Inquiry into the Nature and Causes of the Wealth of Nations* (New York, 1937), pp. 7-8. "Dexterity" is the term actually used by Smith.
4. Ibid., pp. 9-10.
5. The other advantage of the division of labor alleged by Smith (ibid., pp. 8-9) is that of time-saving. Smith claims that much time would be lost if each person had to produce an entire article from start to finish. In making this claim, Smith seems to have underestimated the range of alternatives to division of labor. Thus in the context of his famous example of pin manufacture, the only alternative discussed is a situation in which each person makes an entire pin from start to finish, then produces a second pin from start to finish, and so on. The production of each pin would thus involve the putting down and the taking up—*by each producer*—of a wide varie-

ty of tools within short periods of time, and in the transition from each phase of pin-making to each other phase there would be a considerable loss of time. True enough, there would be such a loss if this were the only alternative to the division of labor. But, as should be plain from the above, there is another. If the people involved in pin-making were to spend equal amounts of time on each aspect of the process, "rotating" jobs every day or two, then the inefficiencies of "putting things down and picking other things up" would be lessened or even eliminated; and we would not have "division of labor" properly so called.

6. Smith makes this point dramatically by speculating that it may even be true "that the accommodation of a European prince does not always so much exceed that of an industrious and frugal peasant, as the accommodation of the latter exceeds that of many an African king, the absolute master of ten thousand naked savages" (ibid., p. 12).
7. Ibid., pp. 734 ff.
8. Smith does not, to my knowledge, explicitly address himself to this particular issue. But the reasons I attribute to him for believing a historical retreat "out of the question" are clearly fundamental to his approach.
9. Smith, *An Inquiry,* pp. 736 ff.
10. Throughout his career, Marx did in fact use the expression "political economy" as essentially synonymous with the contemporary Marxist's expression, "bourgeois economic theory."
11. Marx's actual words are as follows: "In handicrafts and manufacture, the workman makes use of a tool; in the factory, the machine makes use of him. . . . In manufacture the workmen are parts of a living mechanism. In the factory we have a lifeless mechanism independent of the workman, who becomes its mere living appendage" (Karl Marx, *Capital,* Vol. 1 [New York, 1967], p. 422).
12. This theme, concerning the increasing obsolescence of industrial labor, occurs in many places in the Marxian opus. A particularly rich source in this regard is the *Grundrisse* of 1857-1858, *passim.*
13. It is not often enough appreciated that the distinction between mental and manual labor is, for Marx, the functional basis of classes. But this is, in my view, one of his most important insights. See especially Karl Marx and Friedrich Engels, *The German Ideology,* Part One (New York, 1970).

14. Talk of the "abolition of the division of labor" occurs not only in Marx's early work (see, e.g., n. 18 below), but throughout Marx's work—e.g., in such a late work as *Critique of the Gotha Programme*. The phrase, "the abolishing of the division of labor," occurs, among other places, in *The German Ideology*, p. 83.
15. Our language does not have a cliché for "universal development" that includes the feminine gender. This is itself a reflection of social realities.
16. Marx does use expressions like "all-around development" (see, e.g., the *Grundrisse, passim*), "cultivating . . . gifts in all directions" (*The German Ideology*, p. 83), and other cognate expressions; but my own interpretation (to follow) of what such "universality" should be taken to mean is, I think, nonetheless correct.
17. Marx is referring, of course, to a well-known tendency among the Young Hegelians.
18. *The German Ideology*, p. 53.
19. Ibid.

Two Comments
by Harry Braverman

I would like to take this opportunity to comment on two of the many issues that have been raised in the accompanying articles. The same issues have been raised in a number of other reviews and communications on *Labor and Monopoly Capital*. The first has to do with the connections between the subject matter of the book and the women's movement. The second has to do with the consciousness of the working class as a class *for itself*, struggling in its own behalf, apart from its objective existence *in itself*.

I

The authors of "The Working Class Has Two Sexes" generously conclude that "*Labor and Monopoly Capital* . . . makes a major contribution, perhaps unbeknownst to its author, to feminist analysis." Be that as it may, the connection did not come as a post-publication surprise to me. During the earliest period of my research, I became convinced of the importance of recent trends in the working population for the feminist movement. I have been gratified to see that many of the conclusions I had drawn in my own mind have now been drawn by readers, and particularly by women readers.

In comments both public and private, many of these readers have expressed some disappointment that I omitted from my discussion any direct comments on matters of special concern to the women's movement, and particularly household, or non-wage family, labor. Since these readers have all been most understanding of the self-imposed limits of my study, and (as in the two articles contained in this collection which address themselves directly to the subject) have adapted their critiques to these limits, I do not raise this here in order to make an unnecessary explanation. But I would like to add one thing to what has been said on this particular point. Beyond the fact that a consideration of household work would have fallen far outside the bounds of my subject (not to mention my competence), there is also this to consider; that household work, although it has been the special domain of women, is not thereby necessarily so central to the issues of women's liberation as might appear from this fact. On the contrary, it is the breakdown of the traditional household economy which has produced the present-day feminist movement. This movement in its modern form is almost entirely a product of women who have been summoned from the household by the requirements of the capital accumulation process, and subjected to experiences and stresses unknown in the previous thousands of years of household labor under a variety of social arrangements. Thus it is the analysis of this new situation that in my opinion occupies the place of first importance in the theory of modern feminism.

Let me add at once that none of this is said in order to disparage the need for an understanding of the specific forms and issues of household labor, of the working-class family, and of sexual divisions and tensions both within and outside the family. But the unraveling of every complex of social reality requires a starting point, and it is my strong conviction that the best starting point in every case is the analysis of the dynamic elements rather than the traditional and static aspects of a given problem. Thus I have a feeling that the most light will be shed on the totality of problems and issues embraced in the feminist movement, including those of household work, by an analysis that begins not with the forms of household work that have been practiced for thousands of years, but by their weakening and by the dissociation of an increasing number of women from them in the last few decades.

To move to a different, although related, point, which has figured in a number of reviews as well as private comments that have reached me: Baxandall, Ewen, and Gordon raise in this connection my use of the distinction between the *social* and the *technical* or *detailed* division of labor. In common with some other reviewers, they treat this as my own invention, calling it "Braverman's distinction." Actually, as my references to this chapter and my use of quoted materials from Marx should make clear, the entire treatment comes from Chapter 14, especially Section 4 of the first volume of *Capital,* called by Marx "Division of Labor in Manufacture, and Division of Labor in Society." In conection with this topic, there is nothing more important to be studied by any modern reader. On the one hand, it is a brilliant example of Marx's historical method. On the other, it contains in fully developed form Marx's most mature conclusions on the subject of the division of labor, and it becomes ever more mystifying with every passing day how so many can discuss Marx's opinions on this subject as though all he ever wrote on it is contained in the few scrappy paragraphs of *The German Ideology* or other early manuscripts unpublished by him, which represent his first reactions to the problem.

Baxandall, Ewen, and Gordon comment that "Braverman's distinction between the social division of labor and the detail division of labor in capitalist industry is not adequate" for an understanding of the whole of the damage wrought by the divisions of labor in society, and they cite specifically the sexual division of labor. They could not be more right. The distinction in question is adequate only to the purposes for which it was fashioned.

Readers who study Marx's chapter carefully will see how he uses one of his most characteristic tools of analysis: He dissociates the elements of the problem historically specific to capitalism from those generally characteristic of human societies, and treats these not just as continuities, one of the other, but in their polar opposition. From this opposition between abstract social categories and specific social forms in which they are cast in a given epoch of history, Marx works up an analysis of extraordinary penetration.

This does not, however, mean that the analysis is directed toward the clarification of the *sexual* division of labor which originates in the long hunting period of human pre-history, and which requires considerably different tools of analysis. On the other hand, an approach so broadened as to include this problem would require a level of abstraction and generality—

in relation to the division of labor in the factory and similar institutions—so extreme as to make it relatively useless for the latter. It is this, I believe, which dictated the approach taken by Marx and followed by me. What I am trying to emphasize here is that an attempt to combine these two analyses—of the division of labor in modern society and of the most general forms of the sexual division of labor—into a single step would only defeat the object of both analyses and create a muddle all around.

II

Labor and Monopoly Capital has been criticized also for its omission of any discussion of the future of working-class consciousness, although in this case too the critics have, like John and Barbara Ehrenreich in their article in this collection, understood and explained the self-imposed limits of my analysis and thus relieved me of the need to repeat any explanations. Nevertheless, a few further words may be usefully said.

Marxism is not merely an exercise in satisfying intellectual curiosity, nor an academic pursuit, but a theory of revolution and thus a tool of combat. From that point of view the value of any analysis of the composition and social trends within the working population can only lie in precisely how well it helps us to answer questions about class consciousness. Thus I do not quarrel with critics who are anxious to see further progress made in that most important side of the analysis. It was my interest in that very question of class consciousness, in fact, which led to my taking up the entire study in the first place.

When I did so, however, I already had the firm conviction that little purpose would be served by a direct attack on the subject, since it did not appear to me to be in any condition to yield to such an attack. Two major preconditions seemed to me to be lacking. The first has to do with the lack of a concrete picture of the working class, what it is made up of, the trends of income, skill, exploitation, "alienation," and so forth among workers, the place of the working population as a segment of the entire population, etc. I thought that my efforts would best be directed toward helping to fill this gap. I might add that since this is still far from accomplished, I believe that many more essays (along the lines of some of those in this collection) will be appearing in the near future.

The second precondition is considerably more difficult to satisfy. It may be described simply by saying that while social conditions have been changing rapidly over the past half century, and the working class along with them, the class struggle has been in a state of relative quiescence in the United States, Western Europe, and Japan—the countries of developed capitalism for which the analysis must be made. We are therefore lacking in concrete experience, for the most part, of the sort which will indicate the forms and laws of struggle which will predominate in the new social conditions which characterize the epoch of monopoly capital—although we do have some interesting indications from the sixties, of which the French events of 1968 are perhaps the most suggestive. Those who have been wrestling with this problem since the thirties best realize how, in the absence of further concrete experience, discussion tends to degenerate into cliché, apologia, and the repetition of old formulas, and how difficult it has become to say anything new or fresh about it.

It seems to me that a fruitful discussion of the working class as a class conscious of and struggling in behalf of its own interests will begin to revive as two conditions begin to be satisfied: first, as a clear picture of the class in its present conditions of existence is formed by patient and realistic investigation; and second, as experience begins of accumulate of the sort which will teach us to better understand the state of mind and modes of struggle of this class.

I would like to make one further comment, having to do with my own attitude on this subject, since there seem to be some questions as well as some misconceptions on this score. Some readers have concluded, chiefly on the evidence of my description of a process of "degradation of labor," that I myself am "pessimistic" about the future of working-class consciousness. But if readers will take the trouble to compare, they will find that the wording which I have used to describe the effects of the capitalist mode of production on the physical, moral, and mental constitution of the working population differs from